ULTIMATE TRAVEL LIST

THE **500** BEST PLACES ON THE PLANET... RANKED

Introduction

We've all got a list of places that we want to see for ourselves: places friends have enthused about, places we've read about, dreamed about. This is our list. It's the 500 most thrilling, memorable, downright interesting places on this planet – and, what's more, we've ranked them in order of their brilliance. These are the places we think you should experience; there are sights that will humble you, amaze you and surprise you. They'll provoke thoughts, emotions or just an urgent need to tell someone about them.

So, how did we do it? The longlist was compiled from every highlight in every Lonely Planet guidebook. Every attraction and sight that had caught our authors' attention over the years was included. This list of thousands of places was whittled down until we had a shortlist. Then we asked everybody in the Lonely Planet community to vote on their 20 top sights, and with a bit of mathematical alchemy, we ended up with a score for each entry in our top 500 places to see.

Our second edition of Ultimate Travel List contains over 200 new entries: either knock-out new openings, sights that have significantly upped their game since the first edition, or places that have become more relevant to the way we travel now. For this edition we also changed how we calculated the rankings, giving extra points to sights that are managing tourism sustainably.

Each entry gives a taste of what makes that place worthy of a spot in this book, and the See It! section is a starting point for planning a trip and visiting responsibly – you can turn to our guidebooks and lonelyplanet.com for more detailed directions on how to visit every attraction in the top 500.

This is Lonely Planet's Ultimate Travel List.
We hope that it will inspire many more travel wish lists of your own.

Contents 01–99

100–199

200–299

300–399

400–500

North America

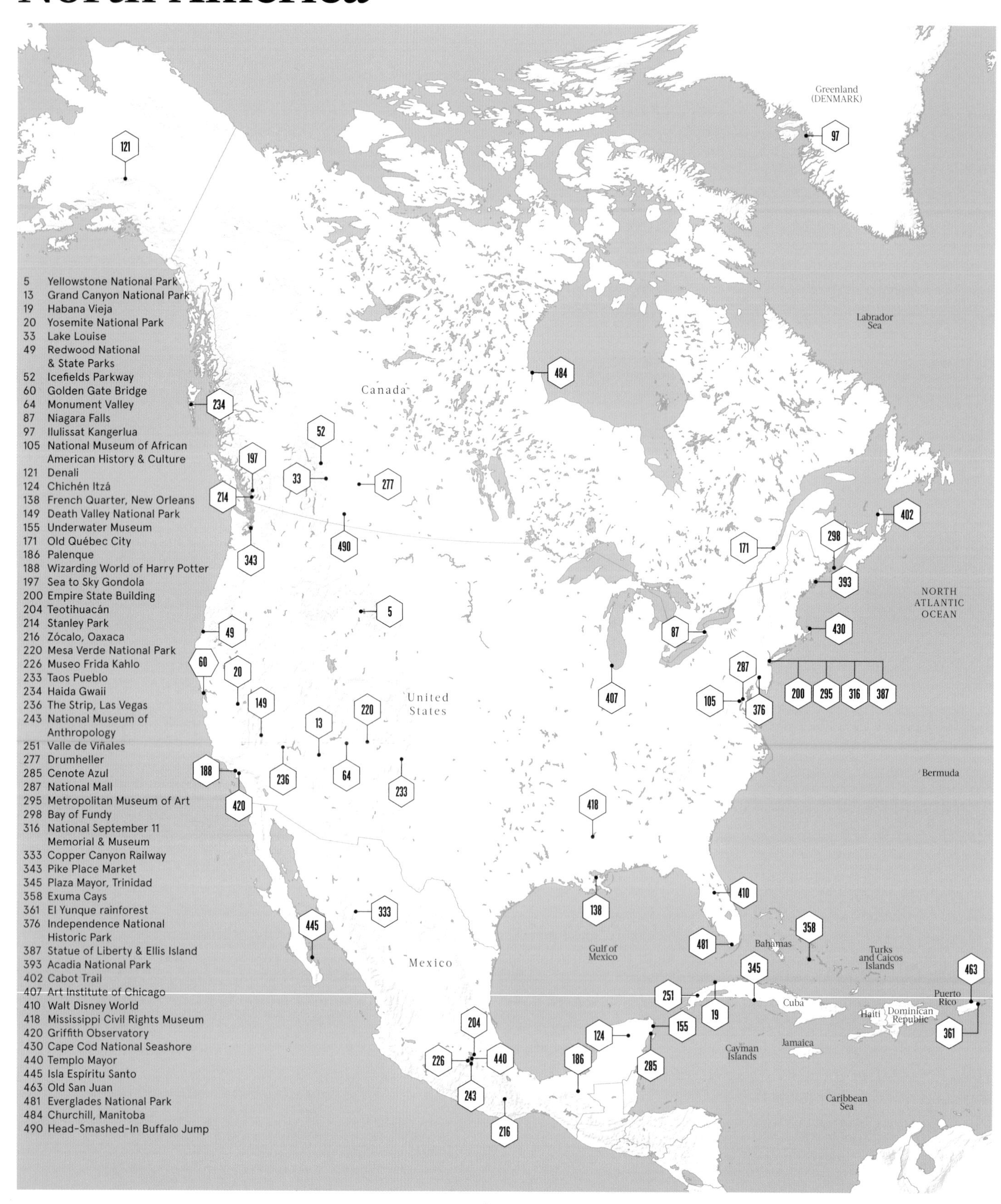

Central & South America

Europe

6 Lake Bled
14 The Alhambra
15 Aya Sofya
21 Wild Atlantic Way
23 Amalfi Coast
27 Acropolis
28 Cappadocia
30 Pompeii
35 Sintra
37 Red Square
40 Lake District National Park
44 Matera
46 Lofoten Islands
48 The Hermitage
50 Colosseum
51 Svalbard
56 British Museum
61 Thermal Baths, Budapest
62 The Cyclades
63 Musée du Louvre
67 Pantheon
68 Auschwitz-Birkenau
71 Knossos, Crete
73 La Sagrada Família
77 Schloss Neuschwanstein
78 Jökulsárlón
79 Musée d'Orsay
82 Wales Coast Path
85 Bay of Kotor
88 The Black Forest
91 Lapland
92 Picos de Europa National Park
98 Siena's Piazza del Campo
102 Giant's Causeway
107 Matterhorn
112 Florence's Duomo
114 Ring of Kerry
116 Jurassic Coast
117 Westfjords
118 Jungfraujoch
125 Plitvice Lakes National Park
126 Versailles
129 Bruges Markt
141 Lycian Way
148 Eiffel Tower
150 Snæfellsnes Peninsula
153 Brecon Beacons
156 The Alfama
160 Grand Bazaar
161 Snowdonia
162 Uffizi
163 Connemara Peninsula
164 Museuminsel
165 Córdoba's Mezquita
168 Ravenna's mosaics
170 V&A
172 West Highland Way
173 Vatican Museums
174 Skógafoss
178 Prague Castle
180 Mont St-Michel
182 Lake Como
183 Real Alcázar
185 St Peter's Basilica
190 Schloss Schönbrunn
192 Pont du Gard
194 Romantic Rhine
195 Vatnajökull National Park
196 Valley of the Temples
208 Charles Bridge
212 Glen Coe
219 Seven Sisters Chalk Cliffs
224 Loch Lomond
227 Stonehenge
228 Tallinn Old Town
229 Swiss National Park
235 Ephesus
238 Skara Brae
241 Dubrovnik
242 Trollstigen
248 Edinburgh's Royal Mile
252 Museum of Broken Relationships
257 Aiguille du Midi
263 Bran Castle
264 Lego® House
265 Brú na Bóinne
267 Roman Baths
270 Sedlec Ossuary
272 Danube Delta
273 Alberobello
276 Meteora
279 Piazza San Marco
280 D-Day beaches
283 Tower of London
291 St Paul's Cathedral
299 Wieliczka Salt Mine
302 Cairngorms' Northeast 250
303 MuseumsQuartier
306 Diocletian's Palace
307 Plaza Mayor, Salamanca
308 Postojna Cave & Predjama Castle
311 Theatre Antique, Orange
314 Christ Church, Oxford
317 Butrint
319 Studenica Monastery
324 Santiago de Compostela cathedral
325 Stari Most
326 Rijksmuseum
329 Louisiana Museum of Modern Art
330 Chernobyl
340 Tate Modern
341 Vestmanna Bird Cliffs
342 Thracian Tomb of Sveshtari
344 Nærøyfjord
349 London's Natural History Museum
352 Hadrian's Wall
353 Gamla Stan, Stockholm
354 City of Arts & Sciences
356 Galleria dell'Accademia
357 Flanders Fields
360 Icehotel
363 Mileştii Mici wine cellars
364 Château de Chenonceau
367 Monet's garden
369 Ffestiniog & Welsh Highland Railway
371 House of Terror
372 Alto Douro wine country
374 Dundee Waterfront
375 Cabo de São Vicente
377 Elbphilharmonie
378 Van Gogh Museum
382 Vasamuseet
385 Arctic Coast Way
389 Sliabh Liag
390 Stirling Castle
391 Brussels' Grand Place
392 Wadden Sea National Park
394 Royal Albert Dock
396 St Davids Cathedral
397 Pulpit Rock
398 Titanic Belfast
399 War Childhood Museum
400 Museo del Prado
401 Lake Geneva
403 Vis Island
411 Teatre-Museu Dalí
412 High Tatras mountains
413 Caminito del Rey
422 Bryggen, Bergen
423 Trinity College
426 Carcassone
428 Kyevo-Pecherska Lavra
432 Schloss Sanssouci
437 St Fagans National History Museum
441 Palace of Parliament, Bucharest
443 York Minster
446 Seyðisfjörður & Borgarfjörður Eystri
449 Markthal, Rotterdam
450 Blue Lagoon, Iceland
451 Kalemegdan Citadel
452 Luxembourg's capital castles
458 Kelvingrove Art Gallery & Museum
459 Perm-36
460 São Bento station
461 L'Atelier des Lumières
462 Castle Howard
467 Pafos Archaeological Site
469 Gallipoli cemeteries
470 Museum of WWII
472 Villa Tugendhat
474 Swarovski Kristallwelten
475 Cité du Vin
476 Portmeirion
477 Bairro of Ribeira, Porto
480 Bauhaus Museum, Dessau
483 Rila Monastery
485 Roman aqueduct of Segovia
488 The Wachau
491 Rynek Główny square
492 Sveti Stefan
497 Cité de la Gastronomie
498 Geysir
Iceland
Norwegian Sea
Faroe Islands
NORTH ATLANTIC OCEAN
Scotland
Northern Ireland
Ireland
Wales
England
English Channel
North Sea
Netherlands
Belgium
Luxembourg
France
Bay of Biscay
Andorra
Spain
Portugal
Balearic Sea
Mediterranean Sea
Alboran Sea

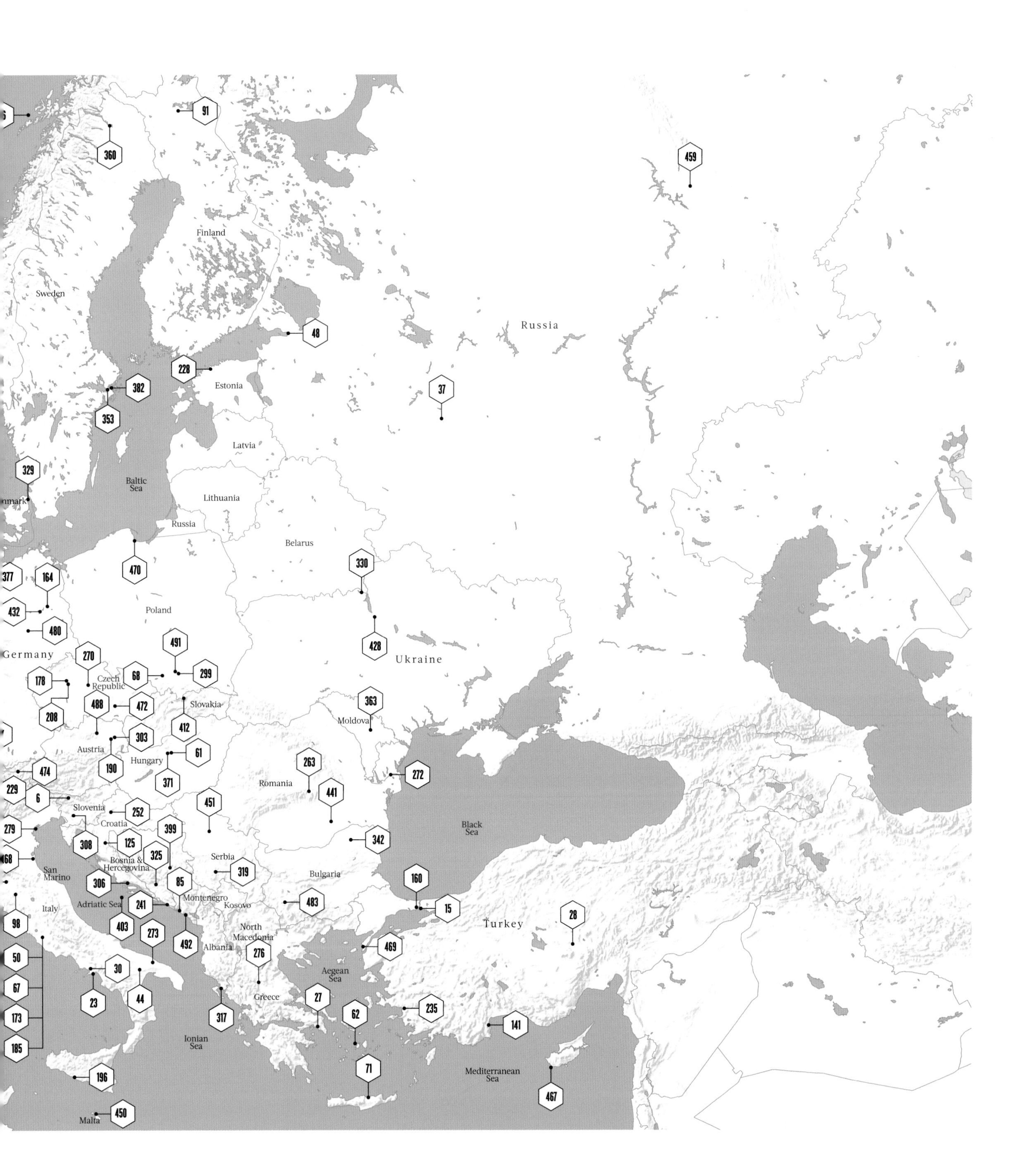
Finland
Sweden
Russia
Estonia
Latvia
Lithuania
Baltic Sea
Belarus
Poland
Germany
Ukraine
Czech Republic
Slovakia
Moldova
Austria
Hungary
Romania
Slovenia
Croatia
Bosnia & Hercegovina
Serbia
San Marino
Bulgaria
Italy
Adriatic Sea
Montenegro
Kosovo
North Macedonia
Albania
Turkey
Black Sea
Aegean Sea
Greece
Ionian Sea
Mediterranean Sea
Malta

Africa

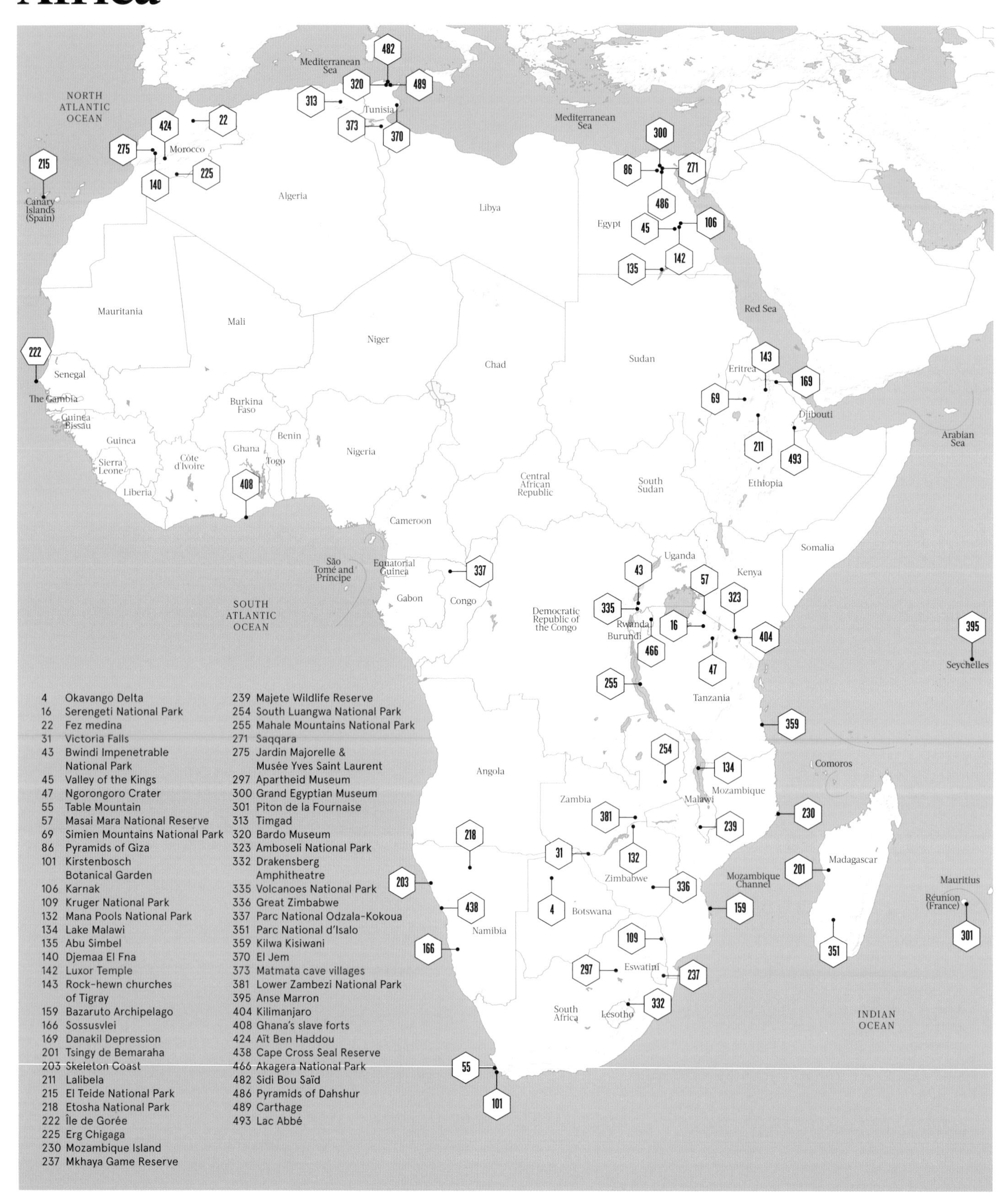
NORTH ATLANTIC OCEAN
Mediterranean Sea
Mediterranean Sea
Morocco
Tunisia
Algeria
Libya
Egypt
Canary Islands (Spain)
Mauritania
Mali
Niger
Chad
Sudan
Red Sea
Eritrea
Djibouti
Arabian Sea
Senegal
The Gambia
Guinea Bissau
Guinea
Sierra Leone
Liberia
Côte d'Ivoire
Burkina Faso
Ghana
Togo
Benin
Nigeria
Central African Republic
South Sudan
Ethiopia
Cameroon
Somalia
São Tomé and Príncipe
Equatorial Guinea
Gabon
Congo
Uganda
Kenya
SOUTH ATLANTIC OCEAN
Democratic Republic of the Congo
Rwanda
Burundi
Tanzania
Seychelles
Comoros
Angola
Zambia
Malawi
Mozambique
Zimbabwe
Mozambique Channel
Madagascar
Mauritius
Réunion (France)
Botswana
Namibia
Eswatini
South Africa
Lesotho
INDIAN OCEAN
4 Okavango Delta
16 Serengeti National Park
22 Fez medina
31 Victoria Falls
43 Bwindi Impenetrable National Park
45 Valley of the Kings
47 Ngorongoro Crater
55 Table Mountain
57 Masai Mara National Reserve
69 Simien Mountains National Park
86 Pyramids of Giza
101 Kirstenbosch Botanical Garden
106 Karnak
109 Kruger National Park
132 Mana Pools National Park
134 Lake Malawi
135 Abu Simbel
140 Djemaa El Fna
142 Luxor Temple
143 Rock-hewn churches of Tigray
159 Bazaruto Archipelago
166 Sossusvlei
169 Danakil Depression
201 Tsingy de Bemaraha
203 Skeleton Coast
211 Lalibela
215 El Teide National Park
218 Etosha National Park
222 Île de Gorée
225 Erg Chigaga
230 Mozambique Island
237 Mkhaya Game Reserve
239 Majete Wildlife Reserve
254 South Luangwa National Park
255 Mahale Mountains National Park
271 Saqqara
275 Jardin Majorelle & Musée Yves Saint Laurent
297 Apartheid Museum
300 Grand Egyptian Museum
301 Piton de la Fournaise
313 Timgad
320 Bardo Museum
323 Amboseli National Park
332 Drakensberg Amphitheatre
335 Volcanoes National Park
336 Great Zimbabwe
337 Parc National Odzala-Kokoua
351 Parc National d'Isalo
359 Kilwa Kisiwani
370 El Jem
373 Matmata cave villages
381 Lower Zambezi National Park
395 Anse Marron
404 Kilimanjaro
408 Ghana's slave forts
424 Aït Ben Haddou
438 Cape Cross Seal Reserve
466 Akagera National Park
482 Sidi Bou Saïd
486 Pyramids of Dahshur
489 Carthage
493 Lac Abbé

Central & South Asia

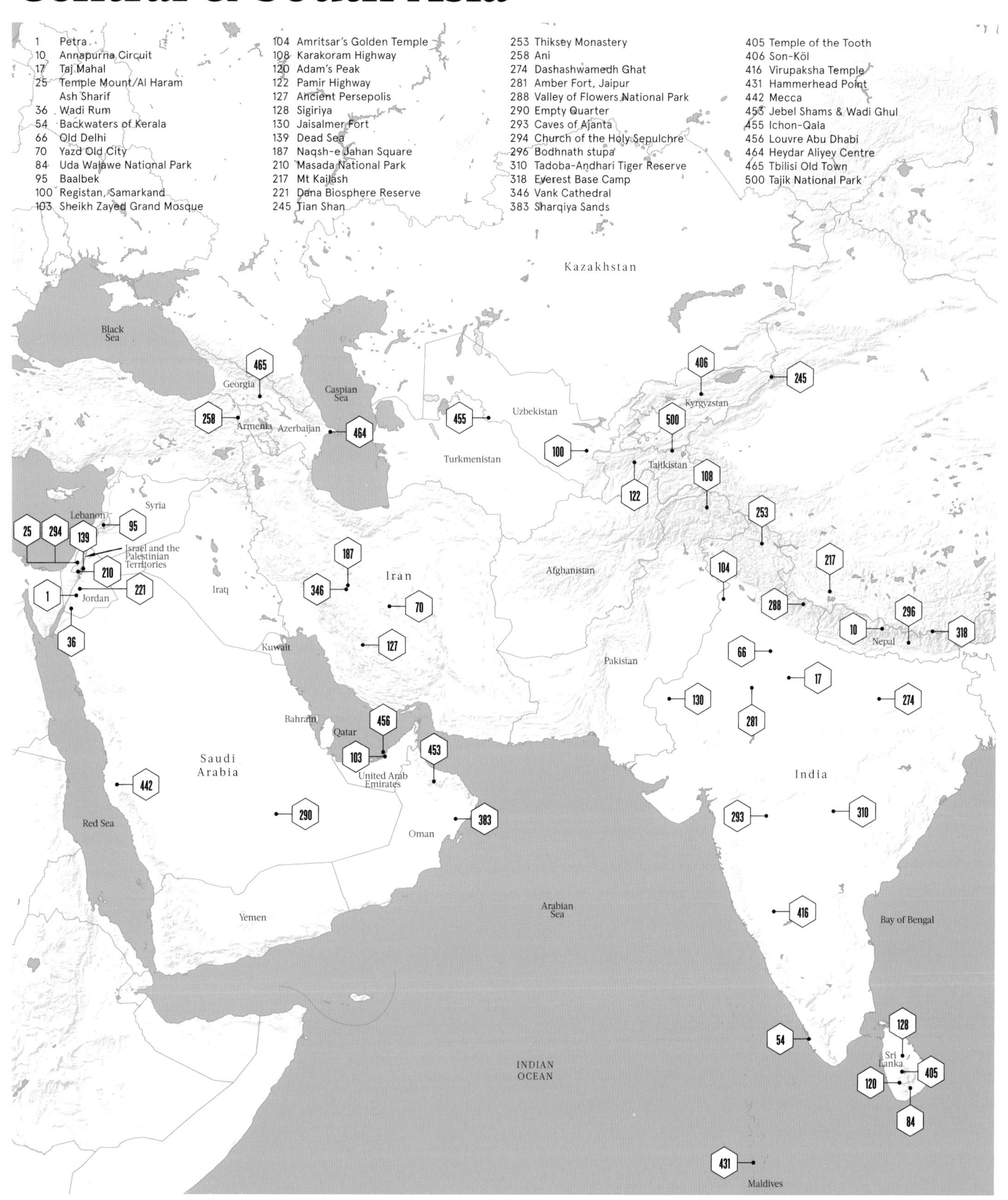

1 Petra
10 Annapurna Circuit
17 Taj Mahal
25 Temple Mount/Al Haram Ash Sharif
36 Wadi Rum
54 Backwaters of Kerala
66 Old Delhi
70 Yazd Old City
84 Uda Walawe National Park
95 Baalbek
100 Registan, Samarkand
103 Sheikh Zayed Grand Mosque
104 Amritsar's Golden Temple
108 Karakoram Highway
120 Adam's Peak
122 Pamir Highway
127 Ancient Persepolis
128 Sigiriya
130 Jaisalmer Fort
139 Dead Sea
187 Naqsh-e Jahan Square
210 Masada National Park
217 Mt Kailash
221 Dana Biosphere Reserve
245 Tian Shan
253 Thiksey Monastery
258 Ani
274 Dashashwamedh Ghat
281 Amber Fort, Jaipur
288 Valley of Flowers National Park
290 Empty Quarter
293 Caves of Ajanta
294 Church of the Holy Sepulchre
296 Bodhnath stupa
310 Tadoba-Andhari Tiger Reserve
318 Everest Base Camp
346 Vank Cathedral
383 Sharqiya Sands
405 Temple of the Tooth
406 Son-Köl
416 Virupaksha Temple
431 Hammerhead Point
442 Mecca
453 Jebel Shams & Wadi Ghul
455 Ichon-Qala
456 Louvre Abu Dhabi
464 Heydar Aliyev Centre
465 Tbilisi Old Town
500 Tajik National Park
Kazakhstan
Black Sea
Georgia
Caspian Sea
Armenia
Azerbaijan
Uzbekistan
Kyrgyzstan
Turkmenistan
Tajikistan
Syria
Lebanon
Israel and the Palestinian Territories
Jordan
Iraq
Iran
Afghanistan
Kuwait
Pakistan
Nepal
Bahrain
Qatar
Saudi Arabia
United Arab Emirates
Oman
India
Red Sea
Yemen
Arabian Sea
Bay of Bengal
INDIAN OCEAN
Sri Lanka
Maldives

East Asia

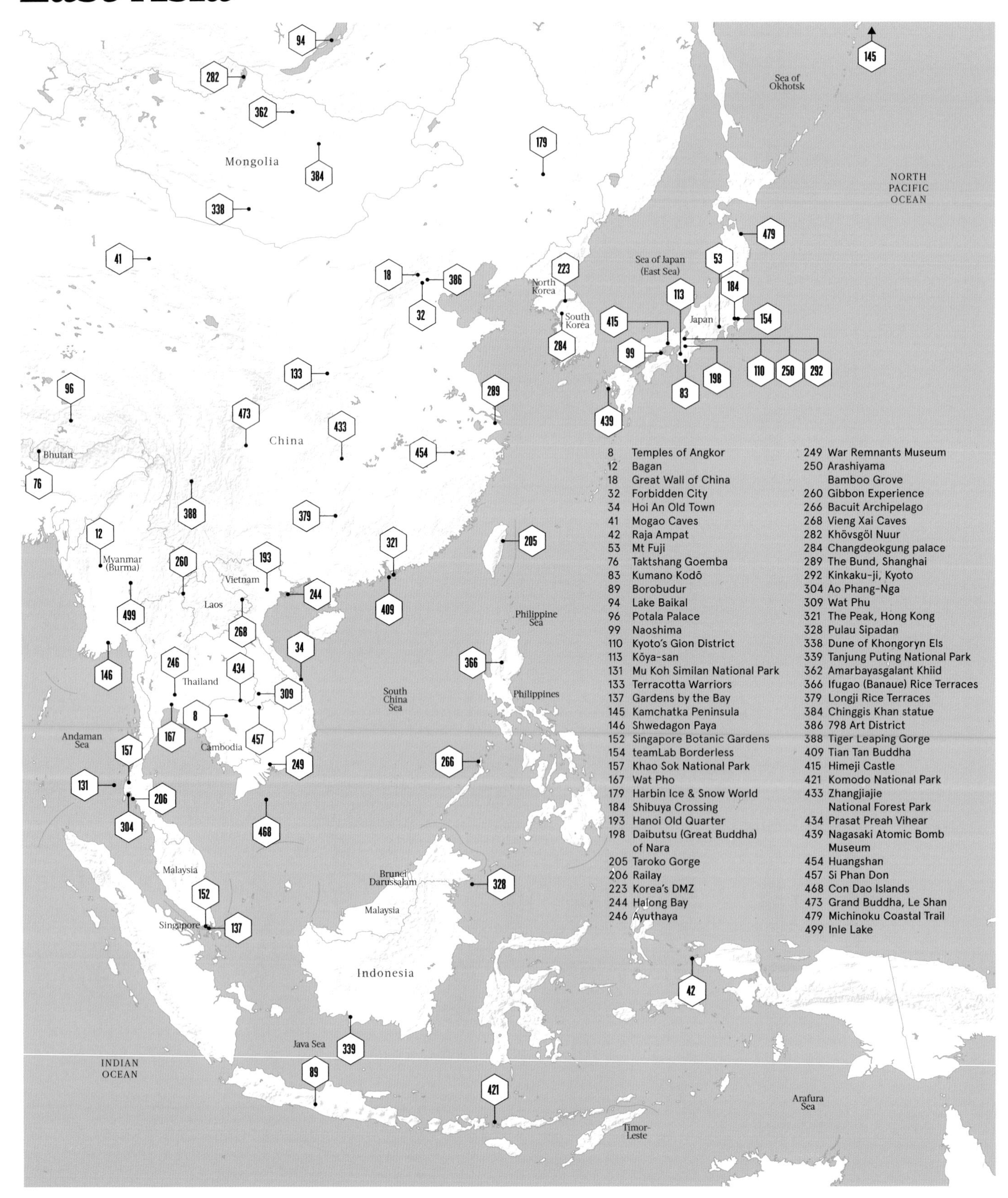

Sea of Okhotsk
NORTH PACIFIC OCEAN
Mongolia
Sea of Japan (East Sea)
North Korea
South Korea
Japan
China
Bhutan
Myanmar (Burma)
Vietnam
Laos
Thailand
Cambodia
Andaman Sea
Philippine Sea
Philippines
South China Sea
Malaysia
Brunei Darussalam
Singapore
Indonesia
Java Sea
INDIAN OCEAN
Arafura Sea
Timor-Leste
8 Temples of Angkor
12 Bagan
18 Great Wall of China
32 Forbidden City
34 Hoi An Old Town
41 Mogao Caves
42 Raja Ampat
53 Mt Fuji
76 Taktshang Goemba
83 Kumano Kodō
89 Borobudur
94 Lake Baikal
96 Potala Palace
99 Naoshima
110 Kyoto's Gion District
113 Kōya-san
131 Mu Koh Similan National Park
133 Terracotta Warriors
137 Gardens by the Bay
145 Kamchatka Peninsula
146 Shwedagon Paya
152 Singapore Botanic Gardens
154 teamLab Borderless
157 Khao Sok National Park
167 Wat Pho
179 Harbin Ice & Snow World
184 Shibuya Crossing
193 Hanoi Old Quarter
198 Daibutsu (Great Buddha) of Nara
205 Taroko Gorge
206 Railay
223 Korea's DMZ
244 Halong Bay
246 Ayutthaya
249 War Remnants Museum
250 Arashiyama Bamboo Grove
260 Gibbon Experience
266 Bacuit Archipelago
268 Vieng Xai Caves
282 Khövsgöl Nuur
284 Changdeokgung palace
289 The Bund, Shanghai
292 Kinkaku-ji, Kyoto
304 Ao Phang-Nga
309 Wat Phu
321 The Peak, Hong Kong
328 Pulau Sipadan
338 Dune of Khongoryn Els
339 Tanjung Puting National Park
362 Amarbayasgalant Khiid
366 Ifugao (Banaue) Rice Terraces
379 Longji Rice Terraces
384 Chinggis Khan statue
386 798 Art District
388 Tiger Leaping Gorge
409 Tian Tan Buddha
415 Himeji Castle
421 Komodo National Park
433 Zhangjiajie National Forest Park
434 Prasat Preah Vihear
439 Nagasaki Atomic Bomb Museum
454 Huangshan
457 Si Phan Don
468 Con Dao Islands
473 Grand Buddha, Le Shan
479 Michinoku Coastal Trail
499 Inle Lake

Oceania & Antarctica

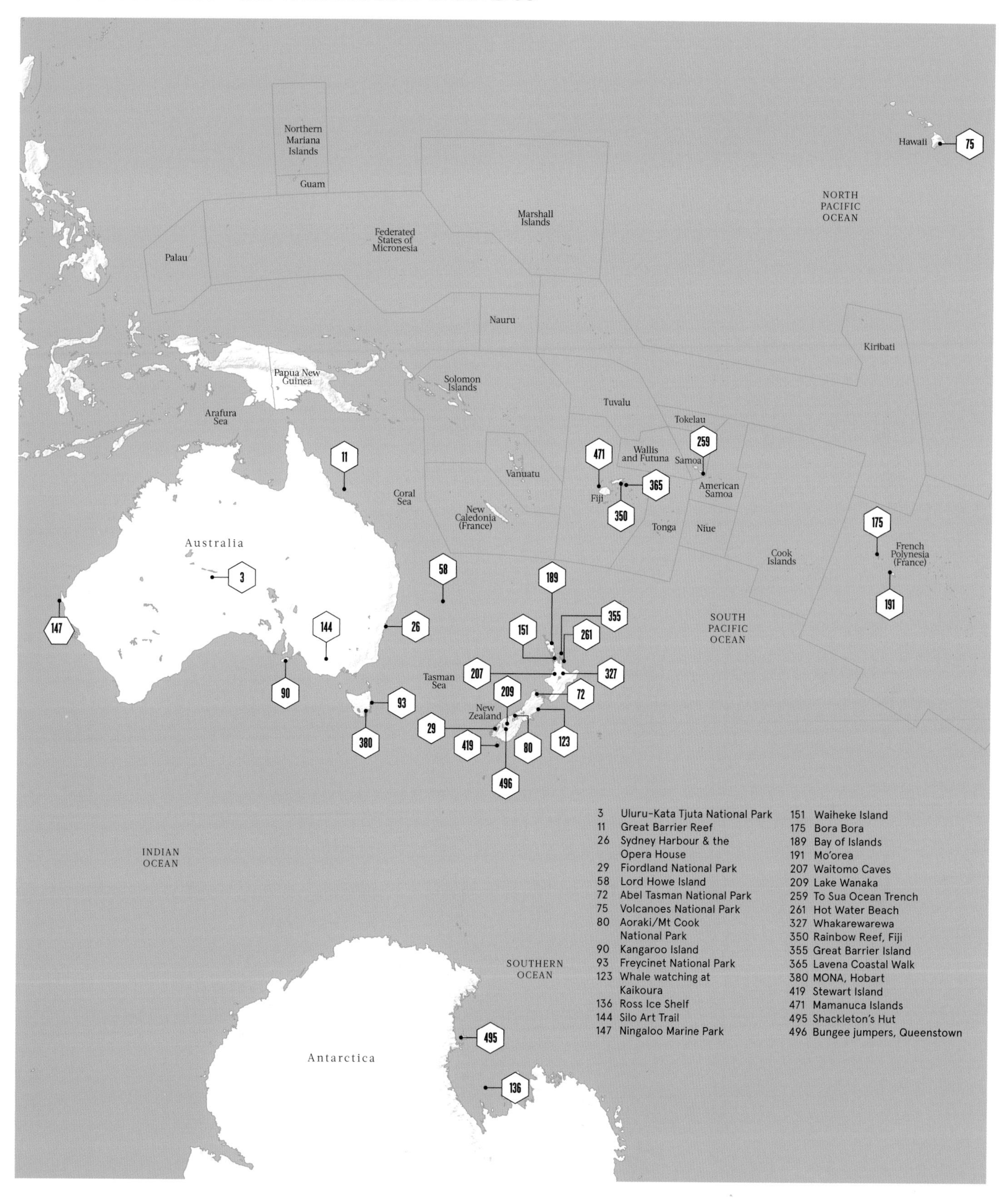

01—99

© kyolshin / Getty Images

Explore the enigmatic 'lost city' of Petra

The astounding Treasury, with its two floors of Hellenic-style columns, offers visitors their first view of this extraordinary monument

JORDAN // Once nearly lost to the outside world, Petra is now one of the most loved places on the planet – and it's our must-see pick. From your first glimpse of the Treasury through the exclamation-point-shaped crack in the canyon you will be lured into the Nabataeans' spell.

Carefully carved into the looming rose-red cliffs, Petra is one of the world's most treasured Unesco Heritage sites and was voted one of the New Seven Wonders of the World by popular ballot in 2007. The sandstone city – spread over some 264 sq km (102 sq miles) – was constructed by the ancient Nabataeans, a civilisation of crafters and prosperous merchants, and made for a grand trade route stop-off between Arabian oases. But generations later, after the city was abandoned, it was nearly erased from the Western consciousness, known only to the Bedouin who made the caves their home.

The difficulty of accessing Petra, secreted away in a series of deep valleys, no doubt helped in its preservation. The iconic Treasury, looming 39m (128ft) high and fluted with two floors of Hellenistic-style columns, is most visitors' first sight of the city, reached after a winding journey through the narrow water-etched slot canyon known as the Siq. The naturally colour-splashed Royal Tombs, a Roman-style amphitheatre and the grand Colonnaded Street lie in wait just around the corner, overseen by the High Place of Sacrifice where the Nabataeans drained the blood of animals to appease their gods.

When Swiss explorer Jean Louis Burckhardt laid eyes upon the Treasury in 1812, he was the first foreigner to witness the 'lost city' in hundreds of years. Today Petra could hardly be more found: it clocks nearly a million annual visitors. However, it's easy enough to rediscover Petra's eternal air of mystery by getting off the beaten track: early mornings and the hours before closing tend to be the quietest and most atmospheric, and the surprisingly frigid and wet winter months of December to February scare off the crowds.

To visit Petra responsibly, take it slow and steady. Sandstone is extremely fragile and easily dissolves if disturbed. Opt for shoes with a light tread and leave hiking poles at home if you can. Reconsider whether you need a ride: some visitors complain about the treatment of Petra's camels, horses and donkeys, though collaboration between the local Bedouin and the Jordanian government has seen the plight of the working animals improve. If you're planning to tackle the 850 steps to the Monastery, pace yourself: erosion can be lessened by taking it slow. Better yet, spend an extra day at Petra and arrive through the 'back door' entrance that drops hikers near the Monastery after traversing an epic mountainside route. Let Petra leave its mark on you instead of the other way around.

SEE IT! ***The town of Wadi Musa is the transport and accommodation hub for Petra. The site can also be reached by minibus from Amman (four hours), Aqaba (two hours) and Wadi Rum (1½ hours).***

02

See the islands that changed the course of science – the Galápagos

ECUADOR // A thousand kilometres from mainland South America, life on the Galápagos follows different rules. Cormorants can't fly. Iguanas can swim. Tortoises don't die (OK, they do, eventually, but some have lived nearly 200 years). Animals, having lived so long without us, don't fear us, and barely blink at a camera snap.

The islands are famed, of course, as the place where Charles Darwin developed his ideas on evolution by natural selection. The process is more obvious here than elsewhere on Earth, since animals that otherwise look alike diverge subtly from island to island, depending on their environment. Nearly 200 years after Darwin's journey on the HMS *Beagle*, naturalists and laypeople alike continue to be astonished by the archipelago's stunning diversity and sheer strangeness. Here you'll find blue-footed boobies performing cartoonish mating dances, colonies of inky-black marine iguanas creeping through the spiky brush, penguins zooming underwater like cruise missiles, puppy-like fur seals lounging on volcanic rocks. The landscape has an off-kilter beauty – brackish lagoons edged with rusty-red carpetweed, bulbous prickly pear cacti (the tortoises' favourite food), copses of fern-draped Scalesia trees, black volcanoes disgorging fresh lava.

The 17 islands are part of the Galápagos National Park. Here, the label 'ecotourism' isn't just greenwashing. The airport runs on wind and solar power and is made mostly from recycled materials. Each municipality has a strict sustainability plan. Tour operators must adhere to clean water and energy rules. Still, it behoves travellers to use companies that go the extra mile. It doesn't take much – a stray plastic bottle, a stowaway mouse – to irrevocably alter this enchanted place.

SEE IT! ***Reach the two main airports from Guayaquil, Ecuador. Cruises are also popular. Once there, independent travellers can use water taxis.***

Opposite: fur seal meets rock crab. Above: the view toward Pinnacle Rock on Bartolome Island. Below: the unusual blue-footed booby

© Ignacio Palacios / Getty Images

Startling, enduring and never failing to impress its thousands of visitors, the Uluru-Kata Tjuta National Park radiates mystery and majesty

03

Take some life lessons from the Anangu at Uluru-Kata Tjuta National Park

AUSTRALIA // At sunset, when its wavy walls blaze gold, Uluru looks like a ship on fire in a desert sea. Rising to 348m (1142ft), the sandstone monolith seizes your eyes from miles away. It's easy to see why it's a sacred site. The Anangu people, the area's original inhabitants, believe it's still home to spirit ancestors like the python woman Kuniya and the hare-wallaby people, the Mala. But it's become an icon to all Australians, a symbolic heart beating in the country's Red Centre.

Until recently, visitors were allowed to summit Uluru. This went against the wishes of the Anangu, who worried about degradation and climber injuries. But in 2017, the park board voted unanimously to ban the practice.

Luckily there are dozens of far more rewarding things to do. Watch dawn paint a rosy blush on the eastern wall. Take in the spectacle of Kata Tjuta, 36 bulbous red rock domes, about 30km (18 miles) from Uluru. Join a ranger-guided walk past red gum tree groves and sacred waterholes. Spy peregrine falcons wheeling over rocky slopes, and spot kangaroos hopping through the spinifex. Ride a bike – or a Segway – on dusty park trails. And learn about the Anangu, who have cared for Uluru since time out of mind. The cultural centre has a fantastic display on *tjukurpa*, the creation period, when mythical beings roamed the land. You can also work with local Anangu artists to create Aboriginal-style dot paintings.

While you watch the light play across the rock's vast surface, consider this: two thirds of Uluru is actually underground. As the Anangu have always known, there's more to this place than the eye can see.

SEE IT! ***Drive from Alice Springs (5½ hours) or fly to the access town of Yulara from major Australian cities.***

© Lauren Pretorius / Getty Images

04

Experience life in the slow lane on the Okavango Delta

BOTSWANA // Not only is this sprawling inland delta of myriad islands and channels one of world's most beautiful wildernesses, but it is also one of Africa's most compelling safari destinations. Each year, the floodwaters of the Okavango River arrive from the Angolan highlands and expand this unique ecosystem to almost 20,000 sq km (7722 sq miles), sustaining vast quantities of wildlife and attracting famed African species galore.

As waters rise under the blue skies, on islands such as Big Chief, prides of lions, packs of wild dogs and herds of buffaloes, elephants and zebra congregate for spectacular shows. In the Savute Channel, leopards have even adapted their predatory nature to hunt for huge catfish, something not seen elsewhere. And the remote nature of the delta has also made it a refuge for rhinos, with more than a hundred moved here from areas in South Africa.

Unlike in many of Africa's other protected areas, safaris in the Okavango are not bound to 4WD wildlife drives. The water levels during the flood allow for exploration by both powerboat and traditional *mokoro* (dugout canoe). Travelling this way is a truly transcendental experience – as the *mokoro* is poled silently through the shallow reed-lined channels you feel immersed in the environment, hearing every bird and animal call, witnessing the mightiest of elephants crossing your path and the smallest of frogs clinging to the grass.

With development and visitor numbers in the delta strictly regulated to protect the environment, the Okavango is one of the most exclusive destinations on the planet. But for adventurous souls who can handle a 4WD and don't mind camping, there are affordable options within the delta's Moremi Game Reserve.

SEE IT! ***The town of Maun on the southern fringe of the delta is the leaping-off point, whether by plane, 4WD or* mokoro.**

Opposite: Okavango life is lived in the slow lane. Above: a leopard surveys the scene. Below: the lilac-breasted roller is just one of the Delta's 444 bird species

© Pixelchrome Inc / Getty Images

© 2630ben / Getty Images

Opposite, from top: the otherworldly eye of the Grand Prismatic spring; Lake Bled's beauty has made it a haven for royalty, presidents and the proletariat

Sniff out geysers and grizzlies at Yellowstone National Park

USA // This place stinks. And it has no manners at all. The rotten-egg whiff taunts your nostrils; your ears are assaulted by a vulgarity of belches, burps and farts. But then, what do you expect when you're exploring the largest geothermal area in the world? Fully half of the globe's entire collection of geysers, mud-pots, fumaroles and other such restless, pungent features are located right here.

More than 500 active geysers spout in Yellowstone's enormous, steaming landscape – Old Faithful being the most famous. And there are hot springs to visit – from the bloodshot eyeball of Grand Prismatic to the travertine shelves at Mammoth. But the wildlife is perhaps the biggest draw. This is like North America's answer to an African safari, although here the Big Five creatures to spot are bison, bighorn sheep, elk, bear and wolf. Watching a herd of shaggy bison warming up by a thermal pool, catching sight of a grizzly bear shambling across a meadow, or visiting in winter to glimpse wolf prints in the snow – all are quintessential Yellowstone experiences.

☛ SEE IT! ***Yellowstone's North Entrance (at Gardiner, Montana) and Northeast Entrance (near Cooke City, Montana) are open all year. Visit in May or October for fewer crowds.***

Emulate erstwhile pilgrims journeying to astonishingly lovely Lake Bled

SLOVENIA // It started with the pilgrims that have come from afar to worship at the photogenic island church for a millennium or more, it continued with the lakeside sojourns of 19th-century royalty and then of former Yugoslav President Tito who had a villa on the shores – and today, tourism to Slovenia's lovely Lake Bled shows no signs of slowing. This blue-green body of water offset by a winsome white church on a green island, abutted by a terracotta-roofed castle and backed by snow-daubed Julian Alps, is Slovenia's premier crowd-puller. Adding to the appeal is its modest size – the lake measures just 2km by 1.4km (1.2 by 0.9 miles), so walking, cycling or driving around its circumference is an easy 6km (4 miles). Hire a *pletna* (gondola), dive beneath the glass-like surface and do, do stay over in one of many beautiful shore-side properties. Slovenia does not rest on its laurels when promoting this paradise and has won praise for eco-initiatives like ecologically designed Garden Village Bled, where a stream snakes through a serene collection of tree-houses and glamping tents.

☛ SEE IT! ***Lake Bled lies less than an hour outside Slovenia's capital, Ljubljana. Visit between June and August, when mild thermal springs warm the lake water.***

05
06

Marvel at the astounding power of Iguazú Falls

ARGENTINA / BRAZIL // Marking the boundary between Argentina and Brazil, the Río Iguazú flows languidly through the jungle before plunging over a basalt ledge with such sudden, furious force that the planet's most awe-inspiring waterfalls are the result.

On the Argentinian side, a boardwalk leads through jungle vegetation replete with butterflies and squawking parrots, passing a series of increasingly impressive falls until finally the Garganta del Diablo (Devil's Throat) comes into view. Sunlight shines through the spray, creating multiple rainbows as the cascading water ricochets up off the river below. The exhilarating cool mist, the high-decibel roar and the thundering vibrations of crashing water remind you in no uncertain terms of the power and splendour of nature.

SEE IT! ***The falls can be viewed on both sides of the border. The nearest towns are Puerto Iguazú (Argentina) and Foz de Iguaçu (Brazil).***

TERRIFIC TEMPLES

↓

With 3595 temples and shrines, Bagan in Myanmar is simply astonishing

page 33

The former capital of Siam, Thailand's Ayuthaya holds a host of ancient temples

page 181

Seek out Laos' Wat Phu, a bevy of tumbledown temples in a dramatic natural setting

page 218

08

Find Hindu heaven at the temples of Angkor

CAMBODIA // A monument to human ingenuity and devotion, the temples of Angkor have an ability to inspire awe that rivals many of nature's contributions to this list's top 10. And so they should – Angkor Wat, the most famous of Angkor's sites, is a representation of Mt Meru, centre of the universe and abode of Hindu gods. Imagine the wonderment of the ancient Khmer as they entered Angkor Wat for the first time: crossing the vast moat, peering up at the 55m (180ft) central tower, gazing at intricately carved bas-reliefs. Beyond Angkor Wat are more than 1000 temples and shrines, including Ta Prohm, its towers gripped by the jungle, and Bayon with its giant stone faces. Avoid the huge crowds with a visit in wet season for (hopefully) that perfect Angkor sunrise view.

SEE IT! ***The gateway city to Angkor is Siem Reap, where you can also arrange locally run cycling, Vespa, food, wildlife and other ecotours.***

Opposite, from top: the mighty waters of the Iguazú plunge over a basalt ledge; Buddhist monks enter the Bayon temple at Angkor Wat

07

08

The world's biggest salt lake is surreally transformed into a vast mirror after rain, reflecting the sky and mountains to create an endless horizon

Strike out on the salt of the earth at Salar de Uyuni

BOLIVIA // Imagine this: you're standing in the middle of the sky. Clouds above. Clouds below. Blue, blue, blue all around. That's what it's like at Salar de Uyuni, the world's biggest salt lake, after rain. The thin layer of water turns the utterly flat salt surface into a vast mirror. The horizon disappears and you seem to float. That's only one of the brain-bending experiences you'll have visiting these 10,582 sq km (4085 sq miles) of salt, high in the Bolivian altiplano. When the lake's dry, the white cracked immensity feels like something from *Mad Max* – a desolate, sun-scorched post-apocalyptic landscape.

In the dry season you can visit the lake's two islands – Isla Incahuasi and Isla del Pescado. Both are craggy husks of land sprouting with cacti (and, increasingly, selfie-taking visitors). See too the geyser fields, hellscapes of boiling mud and sulphur steam. Several of the eerie high-altitude lakes, turned lurid aquamarine from high mineral content, are home to flocks of flamingos. Train your camera on herds of vicuña (a non-domesticated relative of the llama) and the occasional lone culpeo (a type of fox).

At the edge of the lake, the settlement of Colchani is a centre of salt mining, producing some 20,000 tons per year. The place to sleep is the Palacio de Sal, made, obviously, of salt. Walls, floor, ceilings and furniture are made from more than a million blocks of salt – no licking allowed (seriously, that's a rule).

When you finally return to civilisation after several days on the Salar, you'll feel as discombobulated as an astronaut coming home from Mars.

SEE IT! ***Salar de Uyuni is 350km (217 miles) south of La Paz. Multi-day 4WD tours depart from Uyuni; you can book a day trip too.***

© Katie Garrod/ awl-images.com

10

Take a classic teahouse trek around Nepal's Annapurna Circuit

NEPAL // The walk around Nepal's Annapurna massif has long been considered one of the world's great treks. The scenery is mesmerising and the sense of journey is psychologically satisfying, starting in rice paddies and climbing through yak pastures before crossing the mighty snow-bound Thorung La pass. It's everything a trek should be – challenging, majestic and inspirational. And at the end of the day some of Nepal's best lodges offer hot meals, apple pie and pots of milky tea.

The truth is that the trek has suffered in recent years, mainly due to road construction. Molar-rattling roads now parallel the first few days of the trek from Besi Sahar and for much of the western section from Muktinath to Beni. Not all is lost, though, because it's the side trips that make this a truly great trek, notably up to high-altitude Tilicho Lake or to the village of Ngawal, where breakfast terraces offer views of snow-capped peaks just across the valley.

SEE IT! ***The launch pad for an Annapurna adventure is Pokhara, a couple of hours' bus or taxi ride from the start and end points.***

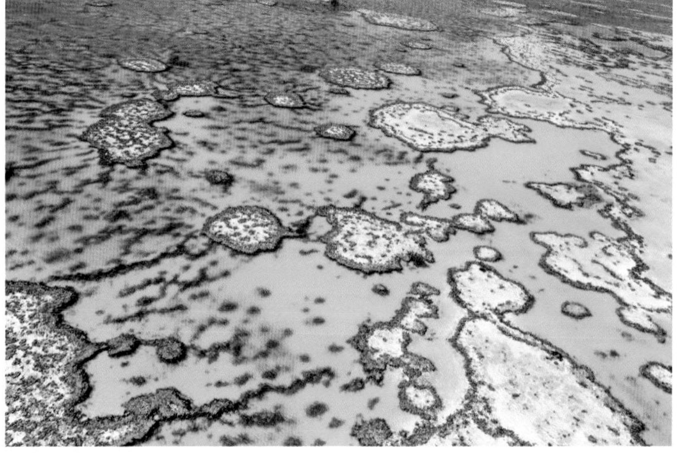

© Edward Haylan / Shutterstock

11

Glide above the kaleidoscopic wonderland of the Great Barrier Reef

AUSTRALIA // The sad truth is this: the Great Barrier Reef is not what it was. Rising sea temperatures have damaged much of the coral in recent years, bleaching its kaleidoscope of colours to a dead-bone white. But don't write off the reef. Its 2300km (1400-mile) structure is still one of the planet's great marvels, home to some 600 kinds of coral and 1500 species of fish, as well as 30 types of whales, dolphins and porpoises and six species of sea turtles. Diving or snorkelling an undamaged stretch means floating above a wonderland of purple sea fans, tangerine-coloured finger coral, and chubby blue sea stars. Hear the 'crack crack' of Hawksbill turtles using their beaks to pry sponges from the coral. Spot clownfish as sweet as Nemo, watch manta rays flap past and maybe – if you're very lucky – catch a slow-moving dugong. The bleached sections of reef could recover in a decade – but only if we throw everything we've got into halting climate change.

SEE IT! ***Popular access points for the reef include Cairns and Port Douglas. Use an operator certified by Ecotourism Australia.***

12

Get a bird's eye view of monumental and ancient temples on the plain of Bagan

MYANMAR (BURMA) // Between the 11th and 13th centuries, a vast city of Buddhist temples and shrines arose in central Burma. Over a millennium later, having endured wars, neglect and devastating earthquakes (the most recent of which, in 2016, damaged some 400 of the 3595 monuments), Bagan stands as Myanmar's most magnificent archaeological site. Don't be put off by the fact that past restorations haven't been conducted in the most pukka fashion: Bagan is a living, breathing site of pilgrimage where many of the temples are functioning places of worship. Awarded World Heritage status in 2019, its prospects are bright. Watching the sun rise or set over the temple-sprinkled plains is as magical today as it ever has been.

☛ SEE IT! *The main hub for Bagan is Nyaung U. Consider visiting during the quieter, cheaper 'green season' between late May and early November. Rent a bike to tour around the monuments.*

© Andrea Bortolameazzi / 500px

Gaze into Earth's mightiest abyss at Grand Canyon National Park

USA // The sheer immensity of the red-rock chasm is what grabs you first: it's a two-billion-year-old rip across the landscape that spans 446km (277 miles) in length and plunges up to 1857m (6093ft) in depth. There is nothing like arriving at the edge and taking it all in – the vastness, the crimson buttes, the long drop down. Sunrises and sunsets are particularly sublime, with the changing light painting the canyon in unbelievably rich hues of vermilion and purple. You won't spend your entire time staring at the hole in the ground, though. You'll also hike on sagebrush-dotted trails and raft the wild Colorado River. April, May and September are the best times to go for good weather and manageable crowds.

☛ SEE IT! *Most visitors access the canyon from the South Rim, about 120km (75 miles) north of Flagstaff, Arizona. The North Rim is more remote, with fewer services.*

© Sumiko Photo / Getty Images

Opposite, from top: the courtyard of the Palacio de Carlos V in the Alhambra; the majestic Aya Sofya, seen from on high

14

Visit the king of all palaces at Granada's Alhambra

SPAIN // Poetry turned to stone; a love-note to Moorish architecture; an opulent recreation of heaven on earth. Choose your metaphor. Usurping India's Taj Mahal and Turkey's Aya Sofya in palatial grandeur, Granada's Alhambra is part palace, part fort, part manicured gardens and part walled refuge from erstwhile Reconquista armies. Still one of Europe's headline acts 700 years after it was built, the red-hued citadel marks what is, arguably, the finest manifestation of architecture to grace the continent since the Romans. The hybridised but harmonious cluster of buildings is crowned by a rich Moorish palace complex conceived by Granada's foppish Nasrid emirs in the 14th century. Embellished with geometric tiles, intricate stucco, placid pools and trickling fountains, no one has come close to emulating their stunning vision since.

SEE IT! *Catch a train here and stay in the Albayzín (old Muslim quarter) just across the valley from the Alhambra. Book tickets online well in advance.*

15

Be dwarfed by the sublime artistry of Istanbul's Aya Sofya

TURKEY // It is appropriate that the Aya Sofya rubs shoulders on this list with the Grand Canyon. Entering the interior of the greatest Byzantine monument has an effect more akin to a vast natural edifice than a building. Above you rises the 'floating dome', described by historian Procopius as 'hung from heaven on a golden chain'. With its concealed pillars and circular plan on top of a rectangular plan, the dome changed the history of architecture and made Emperor Justinian's 6th-century basilica the world's finest building for a millennium.

The finer details of this church converted to a mosque by the Ottomans and to a museum by Atatürk, the founder of the Turkish republic are outstanding too; climb to the upper galleries to see golden mosaics of Byzantine emperors and empresses, and find runic graffiti carved out by 9th-century Vikings in the marble balustrade.

SEE IT! *The Aya Sofya (Hagia Sophia) is in Sultanahmet, Istanbul's old town. There's a separate entrance for the Aya Sofya Tombs, the last resting place of five Ottoman sultans.*

14

15

Opposite, from top: Burchell zebras drink at a waterhole in the Serengeti; preparing for sunrise at India's most famous monument, the Taj Mahal

Feel wildebeests' hooves pounding across the plains in Serengeti National Park

TANZANIA // Each year, Serengeti National Park spawns one of the world's most magnificent wildlife spectacles. It starts on the short-grass plains in the southeastern part of the park between January and March, when more than 8000 wildebeest calves are born each day. In late April up to 1.5 million start 'The Great Migration' – an eight-month round trip towards northern Serengeti and Kenya's Masai Mara in search of better grazing lands. Along the way, the animals may travel well over 1000km (620 miles), dodging the snapping jaws of hungry crocodiles and river currents, watched on by some of Africa's greatest safari species: lion, elephant, rhino, giraffe and buffalo. Predators, especially, are a highlight, with lions, leopards and cheetahs frequently spotted in the open country around Serengeti's central Seronera area.

Even when the migration isn't on, the Serengeti remains a superb safari destination, thanks to large permanent populations of wildlife, notably in central Serengeti and in the park's Western Corridor.

SEE IT! ***Reach Serengeti by vehicle or light aircraft from Arusha in Tanzania. Mwanza is a good springboard for the Western Corridor.***

Find serenity among the crowds at the Taj Mahal, the pinnacle of Mughal architecture

INDIA // The Taj Mahal is so much more than India's most perfect building. Sure, the acres of white marble seem to float weightlessly above the surroundings in perfect symmetry. And the *pietra dura* inlays of semi-precious stones simply stun in their intricacy. It's not even the sublime setting by the banks of the sacred Yamuna River, surrounded by jewel-like formal gardens, that clinches the deal. What truly elevates the Taj Mahal is its story and symbolism.

Built by the Mughal Emperor Shah Jahan as a mausoleum for his favourite wife, Mumtaz Mahal, the Taj has become a symbol of timeless yet lost love. The pathos only deepens when you learn the emperor spent his final years incarcerated in nearby Agra Fort, with just a window view of the Taj to remind him of everything he had lost.

Despite the hordes of visitors the Taj is still a stunning place to visit. There's no other building in India that so perfectly sums up the attitudes and atmosphere of its era.

SEE IT! ***Agra is the main base for Taj tours; to escape the crowds arrive at dawn or book a moonlit visit.***

16

17

Opposite, from top: the Great Wall of China, NOT visible from space, but still impressive; Habana Vieja is arresting and arrested in time

Walk the ramparts of the Great Wall of China

CHINA // One of the world's most inspiring monuments and China's top drawcard, the Great Wall has long held legendary status in the minds of travellers. Though it's a popular myth that the Great Wall is visible from space, its vastness is indisputable: it stretches across 8850km (5500 miles) from the rugged mountains near Beijing to the dune-clad deserts of Gansu. Not one wall, but a series of many smaller segments, the Great Wall was conceived by China's first emperor, Qin Shi Huang (221–207 BC) and construction was carried out for thousands of years by subsequent dynasties. The most recognised brick-tower sections near Beijing were built during the Ming dynasty. These tend to be the busiest with tourists, as they are easily accessible on a day trip. But venture farther west into Shaanxi and Gansu, and you'll find desolate sections of wall made of mud that you can have nearly to yourself.

☛ SEE IT! *Try Mutianyu in January for a chilly but quiet experience, or trek the 'wild' (unrestored) sections (with a guide).*

Lose yourself in the crumbling colonial time trap of dazzling Habana Vieja

CUBA // With life still largely frozen around 1959, year of the Cuban Revolution, Havana is a sun-bleached mass of contradiction, chaos and incomparable beauty. Architecture-wise, the new builds are art deco masterpieces and a standard street bursts with vibrant yet decrepit colonial edifices. Transport? The wheels are mostly vintage, held together seemingly by magic. Culture? The music scene is giant, from jazz notes to rumba rhythms, while Che Guevara still graces billboards like he's today's trendiest A-lister. The Old Town, Habana Vieja, is the beating heart of this bedazzling medley: all of the above at its most intense. Unesco-listed, it's a feast for the eyes but also other senses: the whiff of peso cigars mixed with the bashing sea and black-market petrol, and the burn of rum on your throat. Increasing modernisation will inevitably diminish Habana Vieja's alluring idiosyncrasies. Come now.

☛ SEE IT! *Direct US-Cuba travel remains limited, but many flights arrive from Canada and Europe.*

18

19
P130 430

© Aaron Meyers / 500px

Sunrise lights up Cathedral Peak, one of the bold granite forms in Yosemite National Park – no filter needed

20

Get high on granite peaks in Yosemite National Park

USA // Could Yosemite be the world's most enduring rock star? It's a place where countless waterfalls burst out of the mountains, giant sequoias scratch at the sky, and bears roam the expanses (and often the campgrounds). But it's the rocks that are Yosemite's undisputed royalty.

Everywhere you look, precipitous granite peaks bubble up from the land. Half Dome hovers like a vast stone wave about to break over Yosemite Valley, and the ominous sentinel of El Capitan guards the entrance to the valley. On these rocks, some of the world's most epic climbing stories have been carved. The first ascent of El Cap's legendary Nose, once considered unclimbable, took 47 days. By 2017, Alex Honnold famously scrambled up El Cap without ropes, ascending 914m (2900ft) in three hours and 56 minutes. Hang out in the meadow below and you'll see the ant-size figures of climbers as they grapple with El Cap's ledges.

So whether you come to watch the rock jocks, or to trek to roaring waterfalls and gape from mountaintop viewpoints, be ready to be awed.

☛ SEE IT! ***San Francisco is the nearest major gateway. The park's waterfalls are at their best around May and June.***

© FedevPhoto / Getty Images

© Sabino Parente / 500px

21

Trace Ireland's exhilarating western coastline via the Wild Atlantic Way

IRELAND // Get ready for the ultimate road trip, following this 2600km (1600-mile) signed coastal route from the craggy headlands of Donegal to the inlets and coves of Cork via the Cliffs of Moher. But experiencing the Wild Atlantic Way isn't just about taking in views through sea-salt-sprayed car windows (though those are pretty special). It's about hiking along rugged mountain trails, cantering over powdery white sand, surfing huge Atlantic breaks, spotting puffins and bottlenose dolphins and stargazing in dark sky reserves.

SEE IT! ***It's possible to travel the route by car, bus or bike. Pick up a Wild Atlantic Way Passport and get it stamped along the way.***

22

Time travel through the perpetual pandemonium of Fez medina

MOROCCO // Some 200 years before Marrakesh was even a glint in the eyes of the Almoravid rulers, Fez medina was flourishing as Morocco's cultural and spiritual capital. Today it's a labyrinth of stooped, tangled alleyways that defies mapping. Confounding? Yes. Intriguing? Absolutely. Spice sellers beckon from *funduqs* (ancient inns), hidden squares reveal beautiful tiled fountains, and camel heads swing from butchers' hooks. Souqs, mosques, palaces and the world's oldest university await discovery on foot – this is the planet's largest car-free urban centre.

SEE IT! ***Only official guides are permitted to show tourists around the medina; watch out for touts and 'faux' guides.***

23

© bluejayphoto / Getty Images

Savour la dolce vita on the Amalfi Coast

ITALY // Despite its bucket-list top billing that fills its cascading towns with buzzing summer crowds, the Amalfi Coast will steal your heart. Its sapphire-blue seas, swooping coast road and maquis-laden cliffs are just so damn beautiful. Romans, writers, Hollywood celebrities and VIPs all flock here. They rattle around Sorrento on vintage Vespas, laze like lizards on pontoon beaches in Praia, linger over indecently good lunches in Cetara, and listen to Wagner in Ravello. But the soul of Amalfi resides in the hills along centuries-old mountain trails. It's called the Path of the Gods for a reason.

SEE IT! ***Trains run from Naples to Salerno and Sorrento. From there SITA buses serve towns along an astoundingly scenic route.***

Celine Cousteau's Top Five Places

Celine Cousteau is a film maker, environmental activist and grand-daughter of legendary explorer Jacques Cousteau. She's also an ambassador for the non-profit TreadRight Foundation, created by The Travel Corporation to safeguard people, wildlife, and the planet.

CHILEAN PATAGONIA – The Andes deliver some of the most breathtaking landscapes on earth – they serve up a mixture of glaciers, alpine lakes and forests. It's an amazing place to get closer to earth and sky at once.

MADAGASCAR – Some of the warmest people I have met on my travels are the Madagascans. The locals are welcoming, eager to help and deeply proud of their country.

BRAZILIAN AMAZON – I first travelled to the Amazon when I was nine and have returned many times; most recently to produce *Tribes on the Edge*, an impact film about threats to the indigenous people of the Javari.

FIJI'S CORAL REEFS – Spending time underwater is like meditation, and one of my favourite places to dive is Fiji. The biodiversity and water clarity make it a special place to observe marine life, beautiful corals and colourful nudibranchs.

INDIA – I work with a non-profit called Wildlife SOS in India, campaigning against the use of elephants for manual labour and raising awareness about their welfare. Spending time with these beautiful animals in this vibrant country has been a highlight of my travels.

© Capkin van Alphen / Causecentric Productions

© Philip Lee Harvey / Lonely Planet

A bird's eye view of mountain-framed Machu Picchu, Peru's most famous bucket-list site

24

Wander (and wonder) through Inca edifices at mountain-framed Machu Picchu

PERU // The drama-charged finale to one of the world's most traipsed hikes, not to mention South American tourism's most visited destination and most exported image, the 15th-century citadel of Machu Picchu, perched along an Andean ridge and backed by giant peaks and jungle, is so famous you will have seen it countless times before you arrive via trail or train at its edge. Better than the setting or the century-old Inca edifices of this site, though, is the sense you're seeing something that is still, in this information-rich 21st century, unexplained. There are theories aplenty – royal retreat, alien landing pad – but they remain just that: theories. You can wander here in a liberating knowledge vacuum, forming your own ideas. Admittedly, crowds blunt the beauty, and over-tourism here is real; Unesco has debated taking the attraction off its list and it has tumbled to this list's South American number four. Know that there are other tantalising treks to Inca ruins in Peru: doing those instead will help protect Machu Picchu. Or succumb, as so many do, and see why the hype is still justified.

SEE IT! ***The old Incan capital of Cuzco is the gateway to Machu Picchu. To trek the Inca Trail you must purchase a permit and travel with a registered operator.***

© Nick Brundle Photography / Getty Images

25

Feel awestruck at a timeless spiritual site in Jerusalem

ISRAEL & THE PALESTINIAN TERRITORIES // In the southeastern part of Jerusalem's Old City, an assemblage of sacred architecture has withstood thousands of years of conflict. Jews know it as *Har HaBayit* (Temple Mount), site of the First and Second Temples. Muslims revere the 140-acre plaza as *Al Haram Ash Sharif* (The Noble Sanctuary). Here, the Dome of the Rock (c 691) is orbited by Mamluk-era buildings. South is Al Aqsa Mosque, where up to 5000 worshippers gather for prayers.

SEE IT! ***Arrive early, bring your passport, dress modestly and leave behind accessories with religious significance.***

© Pete Seaward / Lonely Planet

Stroll in the sunshine along Sydney Harbour and the Opera House

AUSTRALIA // Melbourne is charming, but Sydney's got that va-va-voom, thanks mostly to its beauty queen of a harbour. In the daytime, ferries flit to and fro as tourists lick ice-cream cones along Circular Quay. In the evening, the sun sets behind the elegant industrial geometry of the Harbour Bridge, and the lights of Luna Park begin to twinkle. And then there's the splendiferous Opera House crowning Bennelong Point, the city's pride since 1973.

SEE IT! ***The Opera House stages performances from opera and rock to comedy year-round, and offers tours.***

The sunset-hit Parthenon offers a lofty view over the legendary Acropolis complex

27

Behold the most iconic building in the Western world at the Acropolis

GREECE // In the flesh, this glorious gift of sculpted marble backed by bright-blue Greek sky exceeds every postcard or poster you may have seen. The largest Doric temple ever completed in Greece, the Parthenon soars above the city of Athens, columns rising white, cream or gold as the light dictates.

It is the highlight of the Acropolis, an ancient hilltop citadel inhabited since Neolithic times (4000–3000 BC). In 510 BC the Delphic oracle declared it the sole province of the gods and Pericles lavished the temple complex with the best materials, artists, architects and sculptors. Picture it as it was: resplendently coloured buildings and enormous statues of bronze or marble plated with gold and encrusted with jewels. Carved pediments and friezes abounded, many now preserved in the Acropolis Museum.

Walk the luminous marble paths between the Parthenon and smaller temples, like the exquisite Temple of Athena Nike, and then pass the birthplace of theatre – the Theatre of Dionysus – as you descend the hill to the museum. Later, view the Acropolis from afar at night to see it bathed in shimmering gold light.

SEE IT! ***Entry is free on the first Sunday of the month, November to March. Combo tickets cover the Acropolis and six other sites.***

Rush-hour, Cappadocia style, as pre-dawn hot air balloons rise in their hundreds over the 'fairy chimney' rock formations

© Jon Arnold / awl-images.com

28

Peek into the fairytale chimneys of Cappadocia

TURKEY // In this rock-cut Turkish dreamland, surreal views pile on top of quirky historical details until you almost believe the local guides who tell you that *Star Wars* was filmed among these cave churches. But Cappadocia, with its curvy valleys of smooth rock and underground cities, is way more impressive than a Hollywood film. It all began around 12 million years ago with a series of megalithic volcanic eruptions, which produced the ash that hardened into soft tuff rock. Several millennia of erosion formed the sinuous valleys dotted with 'fairy chimney' rock formations, resembling mushrooms, phalluses, camels, Napoleon's hat, the Virgin Mary and anything else you can imagine.

Adding a human touch, Byzantine Christians turned these remote valleys of central Turkey into monasteries. The finest is preserved by the Göreme Open-Air Museum, where the colourful religious frescoes in the fairy chimney chapels look as good as new. You can also tour the upper levels of the underground cities carved by the troglodyte monks, who went to ground there when marauding Islamic horsemen passed through.

Cappadocia was for centuries a farming region, where the holes peppering the fairy chimneys and cliffs were pigeon houses, used to collect the birds' droppings for use as fertiliser. These days, boutique hotels offer rock-cut hammams and cave suites. Days here are filled with singular experiences, starting with a dawn hot-air balloon flight over the wavy valleys and knobbly formations, such as Uçhisar Castle. Explore the valleys on foot or horseback, and ride a scooter to the old Ottoman Greek village of Mustafapaşa. Come sunset, position yourself on a viewpoint or hotel terrace to see how Rose Valley got its name.

SEE IT! ***The closest airports are in Nevşehir and Kayseri, with flights from Istanbul and beyond. It's a 12-hour bus ride from Istanbul.***

Opposite, from top: the towering peaks of Milford Sound; take a walk through Pompeii's evocative ruined streets

Pick your way through primeval land in Fiordland National Park

NEW ZEALAND // On the edge of New Zealand's South Island is a landscape for giants. Cliffs a kilometre high. Jagged mountain peaks wreathed in clouds. Immense alpine lakes. The deep wet forests feel like something from the Jurassic, dense with ferns and mosses. Birds unique to this part of the world roam the tussocks or shelter amid knobbly tree roots – kiwis, crested penguins, flightless parrots. Dolphins gambol in the glacier-carved sounds. Seals snooze in piles on rocky beaches.

For humans, the rugged land offers some of the world's most mind-blowing kayaking and hiking. The three Great Walks – the Milford, Kepler and Routeburn tracks – take you into the heart of the wild and offer the chance to sleep in old-school tramping huts.

SEE IT! ***A good base is the town of Te Anau, two hours south of Queenstown, with flights to many New Zealand and Australian cities.***

Experience the vibrant life and hellish death of Pompeii

ITALY // Step through the *Porta Marina* (Sea Gate) and BAM! you're back in ancient Roman times in a throng of people descending on the Forum, as they must have done 2000 years ago when Vesuvius blew its stack. Deep ruts still mark the roads where once chariots careened, walls sport electioneering slogans, and row after row of houses, shops, spas, temples and theatres (sometimes housing contorted body casts of their fleeing former inhabitants) line the vast, 66-hectare (160-acre) city. There's plenty to explore: the Temple of Isis, which inspired Mozart's *Magic Flute*, the House of the Tragic Poet, the ever-popular brothel with its lewd graffiti, the grassy amphitheatre and the steamy Stabian Baths. To finish, make for the Villa of Mysteries, daubed with blood-red frescoes, from where you can still spy Vesuvius glowering.

SEE IT! ***It's a short train ride from Naples and Sorrento. Visit in spring and autumn and carry water on-site.***

29

30

© Vadim Petrakov / Shutterstock

Double the height of Niagara, and known to locals as 'the smoke that thunders', Victoria Falls is best appreciated from above

31

Feel the thundering smoke at Victoria Falls

ZIMBABWE / ZAMBIA // Known in the local Sotho language as 'Mosi-oa-Tunya' (the smoke that thunders), Victoria Falls is where the Zambezi River becomes a roaring sheet of white as it plummets 108m (354ft) and explodes into foam. On the border of Zimbabwe and Zambia, the falls are set in a landscape of wonders: zigzagging basalt gorges, banks lined with wild date and ivory palms, stands of teak, islets topped with snarls of vine like Medusa's hair. Visitors come to see the water, yes, but also to spot wild megafauna like elephants, baboons, rhinos and hippos, and to catch adrenaline thrills with helicopter rides, bungee jumping and white-water rafting the bucking rapids of the lower falls. The brave (or foolhardy) soak in the 'Devil's Pool', where a natural rock wall forms an invisible underwater barrier between swimmers and the abyss. The 115-year-old Victoria Falls Bridge straddles the Second Gorge of the falls, linking Zambia and Zimbabwe in spectacular fashion. Today, visitors can peer over the edge as they walk over – or underneath! – the bridge.

SEE IT! *You can visit from either side of the falls. Livingstone, Zambia, is a popular base, though the Zimbabwe side has the classic views.*

© Matt Munro / Lonely Planet

32

Find more imperial grandeur than ever in the Forbidden City

CHINA // Beijing's imperial palace served as residence to Chinese emperors, and their hundreds of concubines, eunuchs and staff, for centuries. Despite its name, the Forbidden City welcomes 16 million visitors a year to its grand gates and halls and vast courtyards, which fit together in gradually smaller segments like matryoshka dolls. They comprise a huge museum of glorious imperial treasures, including the emperor's Dragon Throne and Buddhas bedecked with every gemstone imaginable.

SEE IT! ***The Forbidden City celebrates its 600th anniversary in 2020 by opening more sections and halls to the public than ever, including the secret Qianlong Garden.***

© Zhukova Valentyna / Shutterstock

33

Rub your eyes in disbelief at bluer-than-blue Lake Louise

CANADA // Standing next to the serene, implausibly turquoise lake, with the smell of pine trees smacking your nose, the natural world feels tantalisingly close. Finely chiselled mountains surround you and hoist up a glistening glacier. They reflect in the mirror-like water, which is so ethereally coloured you begin to wonder if Mother Nature has gone 21st century and clicked on an Instagram filter. When you're finished gawping, trails climb to alpine teahouses that reward with cakes, hot chocolate and more vistas of a certain lake.

SEE IT! ***Lake Louise is in Banff National Park, Alberta. Shuttle buses travel from Calgary and Banff Town, though a car is easiest for getting around. Go in the early morning.***

© Michele Falzone / awl-images.com

© Taiga / Shutterstock

34

Walk through history in Hoi An Old Town

VIETNAM // With a name meaning 'peaceful meeting place', Hoi An has been an important Southeast Asian port for centuries. Ships from across Asia and Europe used to call, and Hoi An's warehouses were full of the treasures of trade, from silk and porcelain to medicines and elephant tusks. Today the riverside Old Town is a charming mix of ornate temples, merchant houses, French colonial buildings, cafes and shops lining narrow streets trimmed with lanterns. Culinary Hoi An is also one of the best places in Vietnam to take a cooking course (or just eat). This Unesco-listed beauty is no secret, though. Walk the streets in the late evening after most tour groups have gone, rise with the sun for quieter temple visits, and hire a bike to get out beyond the town to the craft villages of the surrounding region.

SEE IT! *Hoi An is on the north-south route between Hanoi and Ho Chi Minh City. The nearest major transport hub is Danang.*

35

Gaze in wonder at the colourful palaces of Sintra

PORTUGAL // If you adore mountaintop castles and candy-coloured palaces, Unesco-listed Sintra is your dream come true. Its pride and joy is the fabulously over-the-top Palácio Nacional de Sintra, with its twin conical chimneys and fanciful interior mixing Moorish and Manueline styles. You'll also want to factor in time to clamber 2km (1.2 miles) up a steep, boulder-strewn hill to the ruined Castelo dos Mouros. When the clouds disperse, the views over the thickly wooded hills to the Atlantic are breathtaking. Also on a high – and with something of an identity crisis – is Palácio Nacional da Pena, the summer house King Fernando II built. It's an architectural frenzy of clashing colours, Gothic facades and onion domes – you'll admire him for it, bless his eccentric pastel-loving heart.

SEE IT! *Sintra is a 40-minute train ride from Lisbon's Rossio station. Take bus 435 from the train station to central Sintra-Vila.*

Whether you explore by camel or on foot, there's lots to discover in the multi-hued desert of Wadi Rum

© Tom Mackie / Lonely Planet

36

Make camp under a million stars in Wadi Rum's majestic desertscape

JORDAN // Deserts don't have to be just sand dunes rippling endlessly to the horizon. They can be a lot more interesting than that, and Wadi Rum is one such desert. It has a wild and raw character that never fails to impress, with its whaleback mountains bursting out of sand-scapes that change ochre, pink and orange as the light turns. If you camp out at night, you'll sleep under a vast sky dusted with more stars than you can imagine.

Wadi Rum is also a surprisingly human landscape. Petroglyphs of gazelles, hunters and herdsmen on its craggy walls mark out ancient byways, while Bedouin tribes still water their goats and camels at its springs.

Foreigners have long been drawn to the region. TE Lawrence made it famous in the West after his time here on campaign during the Arab Revolt, and several desert sites are still named for him. In more recent years, Hollywood has recognised something otherworldly in the landscapes, with Wadi Rum standing in for the red planet of *The Martian*, as well as numerous planets from the *Star Wars* movies.

SEE IT! ***Minibuses run from both Aqaba (one hour) and Petra (two hours) to Wadi Rum Visitors' Centre, where tours can be booked on arrival.***

© Joe Windsor-Williams / Lonely Planet

Opposite, from top: the colourful onion domes of Moscow's St Basil's cathedral; the razor-edged giant dunes of Chile's Valle de la Luna

Find the beating heart and bloody past of Russia in Red Square

RUSSIA // It's hard to think of a patch of cobblestones with more history than Moscow's Red Square. On the south end are the whimsical onion domes of St Basil's Cathedral, built in the mid-1500s on the orders of Ivan the Terrible. To the west, a rust-coloured ziggurat is Lenin's Mausoleum, where the embalmed body of the Soviet leader still lies in state. Beyond are the imposing red ramparts of the Kremlin complex of cathedrals and palaces, many dating to the 1400s. Red Square has been a marketplace, a site of religious ceremonies and of executions, a parade ground, a protest spot and so much more. To stand at its centre is to stand in awe at the sweep of history.

SEE IT! *Three of Moscow's metro lines have stops at Red Square. At night, the Square is theatrically lit and empty of bus crowds.*

Stargaze in the magnificent empty expanse of the Valle de la Luna

CHILE // When NASA tests moon gear, it visits the spectral expanse of the Atacama Desert, the driest place on earth. But there's plenty for earthlings to enjoy. Watching the sunset en masse from atop a giant sand dune in the Valley of the Moon is a ritual for travellers, an inspiring cap to an active day of exploring salt flats, watching bubbling geysers and seeing desert canyons ringed by distant volcanoes. Indigenous and colonial traditions continue in the nearby walled adobe villages and whitewashed chapels. At night, bonfires light up open-air courtyards under the stripe of the Milky Way. Don't miss the stargazing – this ranks among the best places in the world to spy the heavens.

SEE IT! *Stay in the village of San Pedro de Atacama, near the transport hub of Calama with flights to Santiago.*

37

38

Opposite: Temple V. Some of Tikal's pyramids soar to over 70m. Right: sunrise over the Tikal plateau. Below: ornately carved temple stelae

39

Walk through Maya history at the pyramids of Tikal

GUATEMALA // Nowhere in Central America speaks to the greatness of the civilisations that thrived before the earth-shattering contact with Europe in 1492 than Tikal. A city that rivalled Rome in size and aspiration, its ruins today poke out of the jungle plateau of northern Guatemala. Where the noise of a bustling metropolis would have once filled the air, visitors today are greeted by the rustling of vines and the shrieks and booms of the toucans and howler monkeys that call its stones home.

The scale of Tikal can be hard to fathom. The main sites can just about be seen in a day, but the park that describes the old city limits spreads out over a staggering 16 sq km (10 sq miles). Less than 10% of Tikal's buildings have been excavated, while the rest lie beneath a dense forest that has been growing since the city's rulers abandoned it a thousand years ago. New archaeological discoveries still come thick and fast – the development of Lidar, a type of laser radar, in the last decade has revealed the remains of an astonishing 4000 previously unknown buildings in the area.

The architecture that draws visitors today is far from hidden. Steep grey temple pyramids loom over ceremonial plazas; the tallest of these towers over 70m (230ft) high, poking its head high over the rainforest canopy.

Given the size of the park, multi-day explorations allow you to get the most out a visit, and Tikal's host of feathered inhabitants offer joy to even the most casual birdwatcher. It's also possible to overnight inside the park, sleeping to a jungle lullaby and rising early to watch the sun rise from atop an ancient temple. There's no better way to feel closer to the ancient Maya world.

☛ SEE IT! ***Forty miles from picturesque Flores, Tikal is among Guatemala's more accessible Maya sites. You can make it a day trip from neighbouring Belize if you rush.***

40

Be moved to your own poetry in the rugged Lake District

Buttermere is backed by green hills and dramatic peaks – nearby Haystacks was the favourite mountain of the writer Arthur Wainwright

ENGLAND // Paths wind up craggy fells, tree-fringed meres stretch along valleys, and stark ridges drop down to lush meadows. The beautiful Lake District has been celebrated for centuries, in the mystical poetry of William Wordsworth, the charming yarns of Beatrix Potter and the adventure stories of Arthur Ransome. And it's even more popular today – indeed this epic national park, home to England's deepest lake and highest peak – was given Unesco status in 2017.

It was recognised for its combination of natural landscape and human activity, and a trip here offers plenty of opportunities to get stuck in. The region is home to England's first via ferrata at Honister, mountain biking in Grizedale Forest and the chance to scale precipitous paths and tramp across rolling hills. For a more laid-back experience, you can explore literary homes and Roman ruins, spot ospreys and cruise on lake steamers and steam trains. There are hearty, rustic places to stay and eat here – but also luxurious yurt camps and world-class restaurants. And wherever you stay or eat, you're rarely far from a view that will set your heart soaring.

☛ SEE IT! ***Keswick, Kendal, Windermere and Ambleside are the main bases. Trains stop at Oxenholme.***

© sihasakprachum / Getty Images

© Igor Tichonow / Shutterstock

41

Peer at Buddhist wonders and walls in the Mogao Caves

CHINA // The Unesco-listed grottoes that make up the Mogao site outside of Dunhuang in western China represent one of the most important Buddhist art sites on earth. The cave paintings were begun here in 366 AD and reached their zenith during the Tang dynasty, when Silk Road traders and travelling monastics stopped to rejuvenate. In addition to the priceless illuminated texts that were kept within (including the celebrated Lotus Sutra), the walls of the caves house ornate murals depicting Buddhas, bodhisattvas and unique flying apsaras. The treasures within the caves went unknown to the wider world until the early 20th century, when European explorers discovered the site and took many of its priceless contents to live in museums abroad.

SEE IT! ***The grottoes are a few kilometres outside Dunhuang. Buy your ticket (transport included) at the visitor centre closer to town.***

42

Island-hop through Raja Ampat's exotic, thriving seascapes

INDONESIA // The world's most biodiverse and awe-inspiring natural aquarium lies beneath this doozy of an archipelago off the northwest tip of Indonesian Papua. Above the surface, visitors can explore more than 1500 far-flung, jungle-shrouded islands and seemingly infinite white-sand beaches. The reefs below teem with carpet sharks, playful mantas, tiny seahorses and brightly coloured coral heads, for which divers and snorkellers travel upwards of two full days to encounter. The difficulty in reaching the islands, along with a hefty marine-park fee, have helped keep this aquatic paradise pristine, and visitors should honour the place by employing reputable guides, following all rules and keeping those flippers off the coral. At some point, towel off and scale the limestone cliff Wayag, arguably the planet's most Instagram-worthy island lookout.

SEE IT! ***You'll have to fly to Sorong, Papua, and then take a ferry out to the islands.***

Around 500 mountain gorillas live in Bwindi Impenetrable National Park ~ about half the world's total population

43

Spend an unforgettable hour with gorillas at Bwindi Impenetrable National Park

UGANDA // Home to almost half of the world's surviving mountain gorillas, the Unesco World Heritage–listed Bwindi Impenetrable National Park in Uganda is one of East Africa's most famous, not to mention most significant, national parks. Arrayed across 331 sq km (128 sq miles) of improbably steep mountain rainforest, this stunning realm now provides refuge for around 500 gorillas. The Impenetrable Forest, as it's widely known, also happens to be one of Africa's most ancient habitats, having thrived right through the last Ice Age (12,000 to 18,000 years ago) when most of the continent's other forests disappeared. In conjunction with the altitude span (1160m to 2607m; 3805ft to 8553ft), this antiquity has produced an astonishing diversity of flora and fauna, even by normal rainforest standards. And we do mean rainforest; up to 2.5m (8.2ft) of rain falls here annually. Add in 120 mammal species and more than 360 species of bird and you have the near-perfect combination of beauty and rich biodiversity.

SEE IT! ***Getting here from elsewhere in Uganda is best done through a safari operator. Permits should be reserved in advance. Avoid March to May and September to November.***

44

Find a drama-filled past in the depths of Matera

ITALY // The oldest troglodyte town in the world after Aleppo and Jericho, Matera was once the shame of Italy. Set on a limestone plateau, it tumbles into a deep river ravine pockmarked with *sassi* (cave dwellings) once inhabited by poverty-stricken peasants, but is now a Unesco Heritage site. History is so deep here it weaves a spell as you wander a labyrinth of hand-cut stairs and alleys visiting cave museums and cafes, sleeping in candlelit cave hotels, and gazing at churches where frescoed poppies swirl around the skirts of a doe-eyed Madonna.

SEE IT! ***Matera is 63km (40 miles) from Bari Airport and accessible by bus or train.***

45

Explore the Valley of the Kings' burial chambers

EGYPT // Follow the footsteps of treasure-hunters and archaeologists to this valley where Tutankhamun and co. were buried with a booty of valuables to keep the good times rolling in the afterlife. The riches are now in short supply – shipped off to museums worldwide. What remains is the succession of 63 royal tombs. Descend into the warren of chambers, covered in lavish scenes of ancient scriptures, for a pharaonic history lesson like no other.

SEE IT! ***This is the big-hitter of Luxor's west bank. Early birds arriving at 6am get the tombs to themselves.***

46

Be blown away by the beauty of the Lofoten Islands

NORWAY // In a country full of scenic highs, Lofoten takes the Norwegian biscuit. Craggy peaks shimmering vivid green in summer and snow-dusted in winter have a Mordor-like other-worldliness. It's pinch-yourself beautiful, especially in the deep Arctic winter when the Northern Lights come out to play. The main islands – Austvågøy, Vestvågøy, Flakstadøy and Moskenesøy, interconnected by bridges and tunnels – each have sheltered bays, sheep-nibbled pastures and off-the-beaten-track fishing villages totally dwarfed by their backdrop.

SEE IT! ***The Lofoten Islands are 68° north of the Arctic Circle, in Norway. Car ferries operate from Bodø to Moskenes, Værøy and Røst.***

Opposite, from top: Ngorongoro Crater is home to zebra and wildebeest; Catherine the Great's glorious Winter Palace now houses the Hermitage Museum

Glimpse Earth's primeval beauty in wildlife-filled Ngorongoro Crater

TANZANIA // Stand on the rim of Ngorongoro Crater – one of the world's largest intact volcanic calderas – and be faced with the classic visitor's dilemma: should I stay, or should I go? Staying up on the rim is tempting, as the views into the crater – with its ethereal shades of blue and green – become more and more spellbinding the longer one looks. Yet, Ngorongoro's fertile floor is just as bewitching, hemmed in by dramatic escarpments hundreds of metres tall, and covered with swamps, forests, the flamingo-studded Lake Magadi and swathes of savannah grasses. And it is here, on the crater floor, where you can witness unparalleled concentrations of wildlife, particularly of lions and other large predators. If this is not enough, it remains one of the planet's best places to see black rhino in the wild.

SEE IT! *Arusha is the gateway, with vehicle rental also in Karatu. For morning quiet, arrive at 6am or sleep on the crater rim.*

Marvel at the world-class arts and crafts of the Hermitage

RUSSIA // On the banks of the River Neva is the gilded-green Winter Palace of Catherine the Great, now one of the world's most magnificent art museums. The empress commissioned the Hermitage in 1764 to store her private collection; subsequent leaders added to the exhibits, which were greatly enriched in the post-revolutionary period with works seized from the wealthy. Today the six-building complex holds everything from Palaeolithic pottery to paintings by da Vinci and Titian, and from Buddhist icons to Fabergé eggs and modernist sculptures. The Winter Palace itself is part of the art; descending the white marble Jordan Staircase will make you feel like a tsar or tsarina of yore. Indulge in a private tour if you can – it's the only way to visit the fabulous baubles in the Treasure Gallery.

SEE IT! *The closest metro station is Admiralteyskaya. Buy tickets online to avoid the queue.*

47

48

© welcomia / Shutterstock

© Justin Foulkes / Lonely Planet

49

Commune with towering trees in Redwood National and State Parks

USA // The loftiest, most ancient trees on earth jab into the sky of northern California. We're talking *Sequoia sempervirens* – better known as the coastal redwoods – that grow up to 115m (379ft). Walking through a stand of these moss-draped beauties can be a spiritual experience. Velvet humidity rises from the forest floor. The loamy soil springs underfoot. A sepulchral quiet reigns, and you're just a speck next to the 2000-year-old giants. The redwoods sprawl through one national park and three companion state parks, all blazing green thanks to the cool, rainy climate.

SEE IT! ***From San Francisco take a few days to drive Hwy 1 north along the Pacific coast and then Hwy 101 beyond Eureka. May to October are the driest months.***

50

Relive the bloodiest of game shows at the Colosseum

ITALY // A monument to raw power, this massive amphitheatre is Rome's most thrilling ancient sight. Gladiators met here in mortal combat, and condemned prisoners fought off wild beasts in front of baying hordes. Crowds of 50,000 would have poured through the 80 entrance arches, jockeying for a top view into the pit. Guided tours take in the subterranean hypogeum, where the full gore of this Roman game show comes to life. Despite the gruesome realities, there's no denying the grace of the arena, which now glows a peachy pink after a 33-month deep clean.

SEE IT! ***Visit early in the morning or late afternoon to avoid crowds, and avoid queues by booking online or buying tickets at the Palatino.***

51

Play spot-the-polar-bear in the Arctic wilderness of Svalbard

NORWAY // Remote and intrepid, you say? Get yourself over to Svalbard. At 78° north, the archipelago is Europe's largest continuous wilderness and the final frontier before the North Pole. Home to more polar bears than people, it's a place of heartbreaking beauty, especially in the depths of winter as you dog-sled across the frozen tundra, the Northern Lights flickering overhead. Join a snowmobile expedition to discover reindeer-dotted valleys or head to the coast where, with any luck, you'll spot whales, seals and walruses.

☛ SEE IT! *Longyearbyen on Spitsbergen is the best starting point for tours. It's a (bumpy) three-hour flight from Oslo.*

Bill Bensley's Top Five Places

Maximalist hotel designer Bill Bensley is a trailblazer for sustainable hotel design, with projects including Shinta Mani Wild in Cambodia's Cardomon Rainforest. His latest project, WorldWild, is a government-approved sanctuary for mistreated animals in South East China.

01

THE DELGER RIVER VALLEY, NORTHWEST MONGOLIA – We've visited the Delger seven times as the fly-fishing is legendary, and limited to 120 anglers per summer. I averaged a fish every 6.5 minutes this season - all with barbless hooks, of course.

02

EASTER ISLAND – Besides the obvious giant stone heads, this very eerie place is an ancient testament that proves that humans can destroy their own environment and in doing so destroy themselves.

03

SOUTH GEORGIA ISLAND – The fog was thick, but we could smell the guano and hear the cries of the penguins and seal pups miles out at sea. Right on top of the dock we saw the spectacle that is 10,000 penguins in one spot.

04

CARDAMON RAINFOREST, SOUTHERN CAMBODIA – The first time I visited was in the midst of the cool monsoon season, when we spotted several types of kingfisher and hornbill, and heard Gibbons. I fell in love with this rainforest and her residents immediately.

05

CONGO RIVER, NORTH OF KINSHASA – Being whisked up river by speed boat to an unexpected pristine archipelago of pure white sand islands, for a day of memories I will never forget.

© Bensley Design Studios

Motor past glaciers and waterfalls on the Icefields Parkway

CANADA // There are amazing road trips, and then there's the Icefields Parkway. Unfurling for 230km (143 miles) between Banff and Jasper National Parks in the Canadian Rockies, it takes in some of the most mind-blowing mountain panoramas anywhere on the Continental Divide. En route you'll pass cerulean lakes, crashing cascades, mammoth moose and the largest patch of unbroken ice anywhere in North America, the mighty Columbia Icefield. So fuel up, sit back, and let one of the world's great scenery shows unfold.

SEE IT! ***Start in the south at Lake Louise, Alberta or in the north at Jasper. Go June to September to avoid harsh, snowy conditions.***

Harness Hokusai on Mt Fuji

JAPAN // Mt Fuji was the mountain muse for Katsushika Hokusai, perhaps the most famous Japanese printmaker, whose images of imperial Japan still endure. Climbing the perfect cone to the barren crater is a Japanese rite of passage, but you'll have to share the experience, and the reward of watching the sunrise from the top, with a crowd. Other (some might say wiser) visitors are content with views of the mountain from the Fuji Five Lakes region, or glimpses as they relax in the shinkansen between Tokyo and Osaka.

SEE IT! ***Official trekking season is July to August, but in recent years this has been extended into September.***

54

Cruise the languid backwaters of Kerala on a houseboat

INDIA // Kerala is far, far away from the hustle and bustle of India you think you know. Book a houseboat trip here and before long you'll be gliding through a network of emerald-green lagoons lined with lush tropical coconut groves, while tucking into a locally landed fish or a cup of Keralan coffee prepared by your onboard cook. Letting the banks slip past, with excited children waving energetically and parents shyly smiling, it's South India's quintessential and most relaxing experience, worlds away from the clamour of modern Indian life.

SEE IT! ***Most cruises start in Alleppey, 85km (53 miles) south of Kochi. Check your boat has a Green Palm Certificate (for waste tanks).***

Table Mountain towers high over Cape Town, offering excellent hiking and phenomenal views

55

Hike up Table Mountain from the city below

SOUTH AFRICA // While some cities have signature buildings or artificial monuments, Cape Town has a trademark mountain. Watching the clouds accumulate on Table Mountain's plateau and then spill over its sides – a phenomenon known as the 'tablecloth' effect – is a genuinely bankable travel moment. You can catch a cable car up the flat-topped mountain's front face; it slowly rotates en route, giving 360-degree views. However, it's better to earn the panoramic view across South Africa's most vibrant city and the table's rocky bookends, Lion's Head and Devil's Peak. Various hiking paths take you to different vantage points.

The classic climb follows Platteklip Gorge, in the footsteps of António de Saldanha, who made the first recorded ascent in 1503. For the best city views, cross the tabletop from the tip of Platteklip and the upper Cableway station to the mountain's highest point, Maclear's Beacon (1088m/3570ft). If your timing's right, you'll be able to join locals watching the full moon rising above the Hottentots Holland mountains to the east, after hiking through the evening to the 669m-high (2195ft) Lion's Head.

SEE IT! ***To experience another side of Table Mountain, hike up Skeleton Gorge or Nursery Ravine from Kirstenbosch National Botanical Garden, or up Kasteelspoort from Camps Bay.***

56

Find a world of wonderstuff at the British Museum

ENGLAND // When Sir Hans Sloane first put his personal collection of treasures and curiosities on display, he probably had no idea what he was creating. Fast forward 270 years and the British Museum has evolved into perhaps the great treasure house of Europe. Indeed, many of the world's greatest artefacts have ended up in its hallowed halls, a periodic bone of contention for the nations where those treasures originated.

The collection houses such famous heirlooms as the Rosetta Stone (the key to the translation of Egyptian hieroglyphs) and the Parthenon sculptures, alongside an extraordinary collection of mummies and sarcophagi that would put ancient Thebes to shame. Our personal top picks? The Mildenhall Treasure, an astonishing collection of Roman silver found hidden in a farmer's cupboard; and the Lewis Chessmen, allegedly unearthed by a cow on the Scottish island of Lewis.

SEE IT! *The British Museum is close to Holborn Tube; free daily tours are led by great guides. Weekday afternoons are the quietest.*

Safari at its finest in the Masai Mara National Reserve

KENYA // This is the place to experience nature at her most ferocious, extroverted and heroic: from waking to see a herd of elephants lumber around your camp to glimpsing a cheetah whoosh at supersonic speed though the grasses, scattering her prey. The Masai Mara meets every wildlife documentary expectation – a vast tract of savannah, dotted with shady umbrella acacias and divided by meandering rivers. Its openness makes wildlife spotting easy but, even so, the trick is not to get hung up on chasing the Big Five – lions, elephants, buffaloes, rhinos and leopards. Instead let the Mara's primeval performance play out before your eyes: be it a sighting as common as a zebra or as elusive as an endangered African wild dog. New community-owned conservancies on the reserve's northern fringes, while exclusive, are providing fresh habitat and seeing increasing numbers of wildlife.

SEE IT! *Peak season in the Mara is July to October during the wildebeest migration – prices often rocket up during this period.*

© Chaokai Shen / 500px

© Pedro Helder Pinheiro / Shutterstock

58

Go all-exclusive on Lord Howe Island

AUSTRALIA // Some 600km (370 miles) off the coast of New South Wales, this boomerang-shaped island is classic tropical paradise – and it intends to stay that way. Only 400 tourists are allowed on the island at any time, which means booking well ahead for a posh ecolodge and the chance to paddleboard the sapphire lagoon, hike the jungle-shrouded volcano slopes and take a boat around the base of Ball's Pyramid, a volcanic sea spire that looks like a medieval fantasy castle.

SEE IT! ***Qantaslink flights connect Lord Howe with Sydney. Visitors are encouraged to volunteer with conservation projects.***

Take in the peaks and pumas of the Parque Nacional Torres del Paine

CHILE // A symbol of Patagonia, granite spikes thrust high above the windy steppe, this 181,000-hectare (448,000-acre) Unesco Biosphere Reserve is a world-class adventure. Hikers pilgrimage to the base of the 2800m-high (9200ft) jagged torres (towers) of rock, but that's just one stop on a superb hiking circuit through beech forests and tawny grasslands at the tip of the great Southern Ice Field. Wildlife includes the Andean condor, puma and herds of guanacos.

SEE IT! ***Puerto Natales has buses to the park. You must reserve all huts and campsites in advance. Go off-season (not December to February) to avoid the crowds.***

Peer through the fog at the towers of Golden Gate Bridge

USA // San Francisco's signature landmark is a soaring, art deco wonder that stretches nearly 3km (2 miles) over the roiling strait below. So what makes it a must-see over other famous spanners? Perhaps it's the way gusts billow through the bridge cables on foggy days, when its two towers are shrouded in a mysterious cloak with near-zero visibility, or maybe it's the wild, green streak of Golden Gate Park that hugs it, where you'll find locals roller-discoing and drum-circling.

SEE IT! ***Pedestrians and cyclists can cross the bridge on its sidewalks; take Muni bus 28 to get there. For best views from afar, head to Fort Point or Marin's Vista Point.***

© Jonathan Gregson / Lonely Planet

© Ungvari Attila / Shutterstock

© Adrienne Pitts / Lonely Planet

61

Cleanse yourself in ancient thermal baths in Budapest

HUNGARY // You've got to love a city where long, luxurious baths are a civic pastime. In Hungarian capital Budapest, locals and visitors revel in the many public bathhouses, which range from 16th-century Turkish-style domed pools to fin-de-siècle palaces. Choose a co-ed or single-sex (read: naked) spa and prepare for an hours-long soak. The sulphur-scented waters supposedly heal everything from arthritis to asthma. Budapest's bathing tradition goes back to the Celts, with the Romans, Ottomans and Austro-Hungarians later adding their own bathhouses. So bathing here is as much about culture as cleanliness.

SEE IT! *Budapest has 15 public thermal baths. Many are male- or female-only on alternate days.*

Sail azure seas and scale volcanic isles in the Cyclades

GREECE // This is the stuff dreams are made of. Ferry between the jewels of the Aegean Sea – from stars like Santorini and Mykonos, to lesser-known but equally fabulous spots like Milos, with traditional villages and a laid-back way of life. Stay on Sifnos, whose whitewashed hamlets line its central ridge, linked by stone paths. See wild, sunburnt and austere Serifos, perfect for desert-lovers. Then there are the large, easy-to-explore islands, like Paros and Naxos, havens for families and dotted with ancient marble quarries and fallen statues. The upshot: every island has its own feel, and you can't go wrong!

SEE IT! *Take a ferry from Piraeus then ferry between them all. Visit May–June or September–October to dodge crowds but catch the sun.*

63

Search out priceless treasures in Paris' classic Musée du Louvre

FRANCE // While many of the 10 million-plus visitors who annually enter the Louvre head straight to the *Mona Lisa* and *Venus de Milo*, the planet's most visited museum is no two-hit wonder. This is an intricately curated record of human endeavour and expression, housed in a 12th-century fortress transformed into a royal residence in the mid-16th century that's as fascinating as its 35,000 exhibits. The glass pyramid main entrance is now a Parisian landmark in its own right.

☛ SEE IT! ***Be sure to pre-book tickets with allocated time slots online. Admission is free from 6pm to 9.45pm on the first Saturday of each month.***

© MasterLu / Getty Images

64

Yeehaw! Pretend you're in a classic Western movie in Monument Valley

USA // Monument Valley's fiery red spindles, sheer-walled mesas and grand buttes have starred in countless films and TV commercials. The place is a legit celebrity. Its scenic beauty sneaks up on you: one minute you're in the drab middle of sand, rocks and infinite sky, then suddenly you're transported to a fantasyland of crimson sandstone towers thrusting up to 370m (1200ft) skyward. The area straddles the Arizona-Utah border and has long been home to the Navajo people.

☛ SEE IT! ***The famous formations are visible from the 24km (15-mile) dirt road looping through Monument Valley Navajo Tribal Park. April/May and September/October are best temperature-wise.***

© Zhukova Valentyna / Shutterstock

© microgen / Getty Images

65

Straddle the top of the world on Pão de Açúcar

BRAZIL // From the peak of Pão de Açúcar (Sugarloaf Mountain), the city of Rio de Janeiro reveals undulating hills and golden beaches lapped by blue sea, skyscrapers sprouting along the shore. Once seen from atop this absurd confection of a mountain, you're unlikely to look at Rio in the same way again. The ride up is good fun: aerial trams whisk you to the top in two stages. The adventurous can even rock-climb their way to the summit. And if the breathtaking heights unsteady you, what better way to regain your composure than with a caipirinha or *cerveja* (beer) on the pinnacle of the world?

☛ SEE IT! ***Sunset on a clear day is the most rewarding time to climb. If possible, avoid peak hours of 10am to 11am and 2pm to 3pm.***

Simon Murie's Top Five Places

Australian open-water swimmer Simon Murie has a solo English Channel crossing under his belt and broke the Australian record when he swum across the Gibraltar Strait. In 2003 he founded SwimTrek, the world's first open-water swimming holiday company.

HELLESPONT, TURKEY – This is where I fell in love with open-water swimming. It has a bunch of significances going for it, including the fact that Lord Byron was the first to actually cross this strait, in 1810.

NORTHERN SULAWESI, INDONESIA – I could easily pick dozens of Indonesia sites, but my current favourite is in the Bunaken National Marine Park. The reefs are spectacular, and the drop-offs will entrance you.

GALAPAGOS, ECUADOR – Why should divers get to see all the good stuff? The Galapagos allows competent swimmers to get up close and personal with some of our planet's most beautiful wildlife.

DORDOGNE, FRANCE – It's more than 300km (186 miles) from the nearest salt water, but the Dordogne River takes you on an aquatic conveyor belt. It's a landscape of cliffs, medieval villages and castles, and you can swim as much or as little as you like.

ANGTHONG MARINE PARK, THAILAND – In southeast Thailand, the isolated islands of Angthong were the original influence for Alex Garland's book *The Beach*, with towering limestone mountains, thick jungle, waterfalls, hidden coves and inlets.

Immerse yourself in the mesmerising chaos of Old Delhi

INDIA // The hurly-burly of motorcycles and pedestrians, the barrage of noise, the searing aromas and the in-your-face colours: Old Delhi is a visceral assault on the senses. Sprawling around the Red Fort, this medieval maze of narrow lanes and hidden temples is anchored by action-packed bazaars, all seeped in Hindu, Sikh and Islamic history, yet brushed with a veneer of modern Delhi life. It will probably make your head whirl but, boy, is it memorable.

SEE IT! ***The best vantage point is the minaret of Jama Masjid – Delhi's largest mosque, which provides a calm respite.***

Admire architectural perfection at the Pantheon in Rome

ITALY // It seems implausible that the Pantheon could have been in use for 2000 years, but it's true. It was commissioned by Marcus Agrippa between 27 BC and AD 14, and rebuilt by Hadrian around AD 126, and it is the blueprint for pretty much every neoclassical building in the world. Its mighty portico is held aloft by a forest of Corinthian columns. Inside, the gravity-defying dome (the largest in the world) still elicits gasps from awestruck visitors. Through the central oculus a shaft of sunlight streams as if heralding an imminent holy arrival.

SEE IT! ***The Pantheon graces the southern end of the Piazza della Rotonda (also one of Rome's favourite spots for an alfresco dinner).***

Shed tears of remembrance at Auschwitz-Birkenau

POLAND // Auschwitz. The name alone causes shudders. Nearly a million Jews along with many Poles, Roma and others were murdered at this extermination camp during WWII. Now it's a museum and memorial to the victims. Beyond the infamous 'Arbeit Macht Frei' entrance sign are surviving prison blocks housing exhibitions as shocking as they are informative. Nearby, Birkenau holds the remnants of the gas chambers. It's essential to visit both to know the extent and horror of the place.

SEE IT! ***Auschwitz-Birkenau is in Oświęcim, Poland. The easiest access point is the historic city of Kraków.***

© Bente Marei Stachowske / Getty Images

Gelada baboons are only found in Ethiopia – they have a fearsome set of teeth and hair like Jon Bon Jovi

69

Get high on life at Simien Mountains National Park

ETHIOPIA // Take away its endemic wildlife and its enigmatic, ever-welcoming population of highlanders, and this national park would still crack the top 500. It's that beautiful, and its trekking routes are simply that spellbinding.

This hulking, mountainous plateau stands tall above the patchwork plains of northern Ethiopia, its chiselled flanks plunging dramatically from countless precipices. Treks here, which can range from a day to several weeks, skirt past pinnacles, along escarpments overlooking Abyssinian abysses and in and out of valleys quilted in blooms.

Although the altitude often passes the 4000m (13,123ft) mark, the toughest part of treks is not the thin air – it's the simple act of moving on. There is so much to stop you in your tracks: the vistas, the wildlife, the people.

The wildlife, particularly the several-hundred strong troops of gelada monkeys (aka 'bleeding heart baboons'), can often seem out of this world. The gelada's large canine teeth, exposed when yawning or under tension, startlingly resemble something from the movie *Alien*, while its long flowing hair is more comical – think Bon Jovi in the 1980s. Meanwhile, huge bearded vultures swoop low overhead, mammoth-horned walia ibex walk along cliff edges below, and Ethiopian wolves slide through the shadows. Incredible.

Longer treks in these mountains can also be a cultural experience, with rewarding encounters with highlanders who have called this area home for time immemorial. Whether crossing paths with a robed herder on a remote ridge (expect a respectful handshake) or stopping in a tiny homestead to buy a goat or chicken for dinner later, your interactions will be as heartwarming as they are memorable.

SEE IT! ***It's straightforward to organise a trek at the park's headquarters in the town of Debark, which is a 2½-hour bus journey north of the city of Gonder.***

© Leonid Andronov / Shutterstock

© Neil Farrin / awl-images.com

70

Stash the map and wander in wonder at the old city of Yazd

IRAN // Time keeps on ticking in the Unesco-listed old city of Yazd, one of the most ancient settlements on earth, which showcases the architectural heritage that's sadly been lost elsewhere. *Badgirs* (traditional windtowers for ventilation) rise above a labyrinth of sun-dried mud walls and earthen roofs, and among the thousands of age-old dwellings is a clutch of hidden courtyards and teahouses, shops selling crafts and houses converted into atmospheric hotels. Narrow, winding *kuches* (lanes) crisscross the town, and it's worth putting your map away and letting yourself be led by curiosity.

SEE IT! ***Yazd is 689km (427 miles) southeast of Tehran, and is well connected to the rest of the country by planes, trains and buses.***

71

Imagine Minoan royalty gliding through palaces at Knossos

GREECE // Knossos was the majestic capital of the mysterious Minoan people of Bronze Age Crete. This astonishing complex, 5km (3 miles) south of Iraklio, perches among the rocks and forests, and the ruins and recreations include a grand palace, courtyards and richly decorated baths. Wild frescoes were added from 1900 to 1930 when British archaeologist Sir Arthur Evans controversially restored parts of the site. Come see whether you agree with how he's done the place up. Either way, you'll leave with a marvellous insight into this special Aegean civilisation.

SEE IT! ***Avoid the worst of the heat by arriving at opening (8am) or early evening, when it's cooler and the light is good for photographs.***

A group of kayaks drifting down the Falls River, in Abel Tasman National Park. The river is so called because it 'falls' 1000m in 10km (3300ft in 6.2 miles)

72

Small and perfectly formed: welcome to Abel Tasman National Park

NEW ZEALAND // New Zealand's national parks can't half be hard work sometimes. Arduous squelches through ankle-deep mud, frigid river crossings, gut-busting climbs over foggy mountain passes. Then it rains. Welcome to New Zealand.

But it needn't be this way. Near Nelson, at the top of South Island, Abel Tasman is a national park pleasure dome – a place where you can swim, sunbathe and kick back, enjoying all the fun of a beach-bum holiday.

The country's smallest national park is also its sunniest and most popular, luring visitors with golden beaches, sparkling seas, coastal forest and granite cliffs. And its Coast Track is New Zealand's favourite Great Walk, with seaside campsites, communal huts and even luxury lodges en route.

The ultimate adventure, however, is kayaking; friendly waters and easy options make it an unforgettable way to explore Abel Tasman's hidden coves, and cosy up to wildlife such as fur seals, penguins and dolphins.

SEE IT! ***The park's an hour's drive west of Nelson, with air connections from Auckland, Christchurch and Wellington.***

73

Catch a rising star at the Sagrada Família

SPAIN // Groundbreaking Catalan architect Antoni Gaudí left this glorious art nouveau basilica rather as Schubert left his Eighth Symphony – beautifully unfinished. Nearly 100 years after Gaudí's death, work continues to deliver the master's magnum opus, a complicated cluster of soaring spires and elaborate facades that has no earthly precedent. Work started in 1882 and proceeded industriously until Gaudí's death in 1926. Since then, progress has slowed to a crawl, although the decorative towers still rise brick by brick. But the basilica remains the most visited monument in Spain and has, somewhat ironically, already required piecemeal restoration.

☛ SEE IT! ***The Sagrada Família is in the centre of Barcelona. A guided tour is recommended to gain understanding of the site.***

74

Go ape in Parque Nacional Corcovado

COSTA RICA // Welcome to the jungle. Famously labelled by National Geographic as 'the most biologically intense place on earth', Costa Rica's greatest national park is enough to make any would-be Tarzan swoon. Muddy and muggy, this is the last great original tract of tropical rainforest in Pacific Central America. The further into the jungle you go, the better it gets: the country's best wildlife watching, most desolate beaches and most vivid journeys lie down Corcovado's seldom-trodden trails. It's home to Costa Rica's largest population of scarlet macaws, as well as countless other endangered species, including Baird's tapir, the giant anteater and the world's largest bird of prey, the harpy eagle.

☛ SEE IT! ***Corcovado is on Costa Rica's remote southwest Península de Osa. Base yourself at Puerto Jiménez or Bahía Drake.***

Rock into Volcanoes National Park

USA // Hawaii sits directly atop a 'hot spot' far beneath the Earth's crust. It's been active for millions of years, and this park spotlights the fiery landscape that has bubbled up as a result. Hiking trails take you into the thick of the primal terrain, past apocalyptic lava deserts, steaming craters, sulphur smells and ancient petroglyphs pecked into the black flowing rock. But the scene is ever changing, based on the whims of the Madame Pele, the goddess who makes her home in the Kilauea Volcano and who shows off her power at will – most recently in 2018's colossal eruption. Only one thing is for sure in this chaotic, churning place: you should check conditions before visiting!

☛ SEE IT! ***The park is on the Big Island of Hawaii; Hilo is the gateway. August and September are the best months for clear skies.***

Hike to shrines at Taktshang Goemba

BHUTAN // The exclusive and enigmatic Himalayan kingdom of Bhutan is jam-packed with sacred Buddhist shrines and golden-roofed monasteries, but none are more spectacular than Taktshang Goemba. Stuck limpet-like onto a vertiginous cliff, and allegedly attached to the rock face by little more than the hairs of angel-like *dakinis*, it is a magical collection of shrines that seems to grow organically out of the rock. Pilgrims come dressed in traditional robes to view the meditation cave of Buddhist sage Guru Rinpoche, who flew here on the back of a tigress; thus the popular name 'Tiger's Nest'. For hikers, a half-dozen temples dot the cliffs above, offering fabulous scope for exploration.

☛ SEE IT! ***Taktshang is an hour's drive and a three-hour hike from Bhutan's only international airport at Paro.***

© Andrew Montgomery / Lonely Planet

© franckreporter / Getty Images

77

Go castle crazy at Bavaria's Schloss Neuschwanstein

GERMANY // King Ludwig II of Bavaria (1845–86) certainly knew how to pick his locations. His fairy-tale pile, Schloss Neuschwanstein, is a riot of towers and turrets surging up from a sheer wooded hill, with the snow-capped Bavarian Alps backdrop the icing on the cake. It's every inch the childhood dream and indeed the Romanesque-revival schloss played a starring role in *Chitty Chitty Bang Bang* and inspired Walt Disney's *Sleeping Beauty* castle.

Obsessed with German mythology and Wagner's operatic works, King Ludwig enlisted a set designer for the interiors, which include the frenziedly frescoed *Sängersaal* (Minstrels' Hall) and Byzantine-style *Thronsaal* (Throne Room). For the postcard view of Neuschwanstein, walk 10 minutes up to the gorge-spanning Marienbrücke (Mary's Bridge). And bear in mind the castle can only be visited by guided tour.

☛ SEE IT! ***The closest town is Füssen, which has regular buses to Hohenschwangau (for Schloss Neuschwanstein).***

Be spellbound by Jökulsárlón iceberg lagoon

ICELAND // Jökulsárlón is one of many awe-inspiring natural wonders in Iceland but it's the country's highest entry here. Why? How did a 17-sq-km (6.5-sq-mile) lagoon of floating icebergs beat waterfalls and volcanoes and fjords? Easy. On a sunny day, with a bowl of lobster soup in your hands to brace against the icy chill, sit beside this mesmerising freak of nature and watch. Hear icebergs calve off the Breiðamerkurjökull glacier, clink and fizz against each other as they drift, continuously, along the lagoon and out to sea. There. A seal pops up. Another. The stillness punctuated with their splashes. What keeps you here? Is it the ever-changing contrast of floating colours: the travelling shards of glacial blue or milky white on the country's deepest lake? This is one of the most perplexing yet spellbinding sights, and you'll leave simultaneously refreshed and challenged.

☛ SEE IT! ***Hire a car from Reykjavik and make this one of several stopovers on a dramatic Ring Road circumnavigation of the country.***

Prepare to be very impressed in Paris' Musée d'Orsay

FRANCE // Degas' ballerinas, Toulouse-Lautrec's cabaret dancers, Cézanne's still lifes, Van Gogh's self-portraits and Monet's gardens at Giverny (ranked #367 on this list) are just some of the instantly recognisable paintings in France's prized national collection from the impressionist, post-impressionist and art nouveau movements (1848 to 1914). The museum occupies an art nouveau showpiece in itself, the Gare d'Orsay, a former train station, with a quintessential Parisian panorama through its giant glass clockface.

SEE IT! *Reduce queueing by buying tickets online. Entry's free on the first Sunday of the month. The museum closes on Mondays.*

80

Go gaga for glaciers at Aoraki/Mt Cook National Park

NEW ZEALAND // At 3724m (12,218ft), New Zealand's tallest mountain cleaves the sky like a knife. In the centre of the South Island, Aoraki/Mt Cook is surrounded by 18 other 3000m-plus (9842ft) peaks, their slopes cloaked in glacial ice. Mountaineers come to tramp, ice climb and bivouac in alpine huts, while the less athletically ambitious take boat rides amid the glaciers of Tasman Lake and sleep in the historic Hermitage Hotel.

SEE IT! *Mt Cook Village has some (pricey) amenities for visitors; catch intercity buses from the nearest town, Twizel, 65km (40 miles).*

Watch the wetlands of the Pantanal burst into wildlife activity

BRAZIL // These twinkling blue-green tropical wetlands sprawl across some 200,000 sq km (77,220 sq miles). They're home to 400 fish species, 1000-odd bird species and an illustrious list of animals, including capybaras, giant river otters, maned wolves, pumas, tapirs, anteaters and the world's most flourishing jaguar population. In the dry season from July to November, the full menagerie descends upon the limited water in one of the most colourful concentrations of wildlife anywhere on Earth.

SEE IT! *Tours here mostly run from the cities of Campo Grande to the southeast, Corumbá to the west, and Cuiabá to the north.*

82

Take in a country's entire coast on the Wales Coast Path

WALES // This little Celtic country's biggest hit on our list is the long-distance path that traces the entirety of its coastline, the 1400km (870-mile) Wales Coast Path. As a result of its completion in 2012, linking up already successful trails such as the Pembrokeshire Coast Path and Anglesey Coast Path in one wondrous sea-hugging route, Wales became the world's first country to have a walking trail all along its coast. Highlights on such a breathtaking trek are harder to pick than the path is to walk. But Wales is a candidate for the most densely castled place on earth, with fortresses standing sentinel along a seaboard featuring some of the most beautiful beaches and cliff scenery anywhere.

SEE IT! ***The path runs between Chepstow in Monmouthshire (in South Wales) and Queensferry in Flintshire (in North Wales).***

© Visit Wales 2016

Track the paths of gods on the Kumano Kodō

JAPAN // From ancient times the Japanese believed the wilds of the Kii Peninsula to be inhabited by gods. To hike through here was an act of worship of nature that attracted ascetics, nobles and retired emperors. Today people from all walks of life take to the World Heritage pilgrimage route known as the Kumano Kodō, but even the modern traveller can achieve something of a spiritual experience here. The Kumano Kodō is not just one route but a network that criss-crosses the peninsula, leading through groves of towering pines, across farmland and along the coast, taking in grand shrines, onsen towns and waterfalls – including Nachi-no-taki, the tallest waterfall in Japan.

SEE IT! ***The main access towns of the Kii Peninsula are connected by train to Osaka. Spring and autumn are the best times for hiking.***

© Jonathan Stokes / Lonely Planet

84

Spot wild elephants on a safari in Uda Walawe National Park

SRI LANKA // The highest of Sri Lanka's offerings on this list, Uda Walawe in the island's south is one of the world's best places to see elephants up close in the wild – around 600 roam the park's grasslands and forests in herds of up to 50, often towing cute babies. There are also lots of deer-like sambar, wild buffalo and jackals, as well as hard-to-spot leopards, and the surrounding marsh and wetlands offer fabulous birdwatching opportunities. The nearby Elephant Transit Home (supported by the Born Free Foundation) offers a bonus chance to watch staff feeding orphaned juvenile elephants, as they undergo their rehabilitation process.

SEE IT! ***Most people visit from resorts on the south coast or Ella. January to March are the best months to come.***

FANTASTIC FJORDS

Iceland's remote Westfjords region promises awesome fjords without the crowds

page 105

Overlooking Norway's Lysefjord, Pulpit Rock offers the ultimate fjordland view

page 266

Hikeor bike around Iceland's spectacular Seyðisfjörður and Borgarfjörður Estri

page 297

85

See the beauty meter climb in the Bay of Kotor

MONTENEGRO // Geologists may quibble over whether the Bay of Kotor is the only Mediterranean fjord, but with a landscape this charismatic, who cares? In the many folds of the bay, lavender-grey mountains tumble down to a fringe of olive and pomegranate trees above the opal sea. Kayaking around rocky coves or sailing to islands topped with monasteries, you'll find the surroundings gobsmacking. The walled town of Kotor seduces with cafe culture, cobbled alleyways and Venetian loggia, while the seaside hamlet of Perast charms with its baroque palazzi fragrant with wild fig. Escape summer crowds on a trek into the craggy hinterland.

SEE IT! ***Kotor town is a perfect base for exploring the bay, or take a day trip from Dubrovnik in neighbouring Croatia.***

Opposite, from top: some 600 elephants roam the grassland and forest of Uda Walawe; Our Lady of the Rocks, an artificial island in the Bay of Kotor

84

85

© sculpies / Getty Images

86

Access the last of the ancient wonders at the Pyramids of Giza

EGYPT // Bah! The Pyramids just scrape into the Top 100? Khufu is probably spinning in his tomb. Well, the world's oldest tourist attraction and the only surviving member of the Seven Wonders of the Ancient World has nothing to prove. Sitting on the edge of Cairo's sprawl, these three megalithic mausoleums for Old Kingdom Pharaohs Khufu, Khafre and Menkaure, guarded by the lion-bodied Sphinx, comprise one of the most recognisable sights on the planet.

We'll admit, plenty of other world-class sights make a better fist of explaining and showcasing the wonders on offer, plus the Giza Plateau is a mega-magnet for would-be guides and hawkers. But all that is forgotten once you clamber into the belly of the Great Pyramid and head down the claustrophobic shafts, enclosed by 2.3 million stone slabs, each placed by the ancient Egyptians' engineering ingenuity. Top 100 or not, the 4500-year-old Giza Pyramids will still be here when you and this list are gone. And crowds will still be flocking to see them.

☛ SEE IT! ***The Giza Pyramids are 9km (6 miles) from central Cairo. Tickets for pyramid interiors are only sold at the main entrance.***

© Igor Sh / Shutterstock

Be soaked in water and mini rainbows at Niagara Falls

CANADA // Niagara might not be the highest waterfall in the world, nor the widest, but it's undoubtedly the most well known. In terms of volume, it's one of the greats – thousands of bathtubs of water plummet over the brink every second. Great plumes of icy mist rise like an ethereal veil as the cascade hits the river below. Rainbows pop up and pierce the haze. But you don't just admire this view from afar. Instead, you sail right towards it in a little boat, getting sprayed, soaked and deafened by the thundering water. The falls forms a natural rift between Ontario and New York State: on the US side, Bridal Veil Falls and American Falls let loose. On the Canadian side, the grander Horseshoe Falls plunge downward. Get a room with a view and you'll be unable to tear yourself away.

☛ SEE IT! ***Niagara Falls is two hours from Toronto, Ontario or 40 minutes from Buffalo, New York. The prime viewpoint is Table Rock (Canadian side). Beat crowds by arriving early.***

© Juergen Wackenhut / Shutterstock

88

Tread softly through the Black Forest's sylvan wonderland

GERMANY // Spreading in all its glory across Germany's southwest, the Black Forest is something of a misnomer – it's actually more green than black, unless you turn up on a snowy winter day. Delve into its remotest reaches and you'll find glacier-carved valleys, where spruce forests rise up above stout timber farmhouses. Here you'll find scenes ripe for a Grimm bedtime story, as well as a welcome dose of peace and quiet.

Slow touring is the way to go: either on foot on the well-marked hiking trails, or on the buses and trains joining the dots. Venture to ridiculously pretty Gengenbach, Triberg (home to the world's biggest cuckoo clock and Café Schäfer, which uses the original recipe for Black Forest gateau), and feel-good Titisee, with its inviting bottle-green lake.

☛ SEE IT! ***Well-connected bases for exploring include eco-aware Freiburg and Baden-Baden. Offered free when you stay overnight, the Konus guest card covers local public transport.***

89

Behold Buddha's mega monument of Borobudur

INDONESIA // If you've seen travel shots of Indonesia, you probably already know what Borobudur looks like. Along with Cambodia's Angkor Wat and Myanmar's Bagan, Java's Buddhist temple complex makes the rest of Southeast Asia's spectacular sites seem almost incidental. The monument's tiered quadrant of bell-like stupas has survived eruptions, terrorist bombs and the 2006 earthquake to remain as beautiful as it must have been 1200 years ago. Nearly 1500 narrative relief panels illustrate Buddhist teachings and tales, while 432 Buddha images sit in chambers on the terraces. Locals call the surrounding countryside the garden of Java, so beautiful is the landscape of green rice fields and traditional *kampong*, all overlooked by soaring volcanic peaks.

SEE IT! ***The gateway town of Yogyakarta is a 1¼-hour bus ride. The site is quietest and most photogenic between sunrise and 6am.***

© Philip Lee Harvey / Lonely Planet

90

Go mad for marsupials on Kangaroo Island

AUSTRALIA // It feels like being in a children's book, this island, where kangaroos are more common than cars, tiny blue fairy penguins waddle around the cliffs at night and that 'crick crick' sound turns out to be a little pink-nosed echidna nibbling on the undergrowth. A road trip takes you from delight to delight: stalactite-dripping caves, a sugar-white beach dotted with lolling sea lions, lavender and honey farms straight out of Provence, giant sand dunes you can ride down on a toboggan. In the evening, sip Shiraz from one of the island's dozen wineries and devour plates of local marrons, a freshwater crayfish, alongside salads topped with cheese made from the milk of the sheep grazing just off the patio.

SEE IT! ***Reach the island by ferry from the South Australian mainland or by plane from Adelaide or Melbourne (in season).***

© Andrea Izzotti / Shutterstock

Monty Halls' Top Five Places

Marine biologist, global expedition leader and TV presenter Monty Halls is best known for the BBC series 'Monty Halls' Great Escape', in which he lived and worked in remote parts of the UK with his dog, Reuben.

BANTHAM BEACH, UK – This wide sweep of white sand, with the mysterious Burgh Island just offshore, is twenty minutes from my house. It's where I go to reconnect and relax, and holds only good memories.

THE CORAL COAST HIGHWAY, AUSTRALIA – An extraordinary ribbon of road that tracks the coastline of Western Australia, it crosses a land with one of the oldest surviving human cultures on earth, and ends at Ningaloo Reef with an encounter with whale sharks.

ISABELA ISLAND, THE GALÁPAGOS – I lived here with my family while we were shooting a series for Channel Four. There really is no place like it; a great volcanic island inhabited by some of the world's most unique animals.

SOUTH AFRICA – A wild, contrasting land, with a coastline tracked by the cold Benguela current to the west, and the warm Agulhas current to the east, the interior's history and heritage is set amid some of the most spectacular landscapes on earth.

DARTMOUTH, UK – I live here for a reason. Bill Bryson said that 'The English countryside is the most reliably beautiful place on earth' and when you walk along the banks of the River Dart, you can see precisely what he means.

Go walking in a real-life winter wonderland in Finnish Lapland

FINLAND // Even if you have no intention of sitting on Santa's knee, Lapland is pure Christmas card stuff. Arriving here in winter is like stepping into a snow globe, with spruce forests bristling above frozen lakes and reindeer-driven sleighs whisking you through snow that lies deep and crisp and even. The bluish twilight of the polar night is the best time to glimpse the Northern Lights, perhaps after you've explored on snowshoes, cross-country skis, dog-sled or snowmobile, or post-sauna in your little log cabin. Santa's HQ, Rovaniemi, is a natural starting point, with every kind of winter activity imaginable. But for a flavour of the remote wonders of Lapland, strike north to Utsjoki on the Norwegian border, where the indigenous Sámi keep ancient reindeer-herding traditions alive.

☛ SEE IT! ***The major gateway to the region is Rovaniemi on the Arctic Circle, where the airport is located. Alternatively, you'll touch down in Ivalo, further north.***

© Miguel Castans Monteagudo / Shutterstock

Scale the heights of Spain's lofty Picos de Europa National Park

SPAIN // The first of Spain's 16 national parks to be established (in 1918), Picos de Europa consists of a crinkled web of glaciated peaks and alpine meadows that straddles three regions of northern Spain (Asturias, Cantabria and Castilla-León), extending over a trio of mountain massifs. As an important pocket of biodiversity rising behind Spain's populated north coast, the park is isolated enough to harbour tiny enclaves of Cantabrian brown bears and Iberian wolves and is blissfully quiet in the winter. This is memorable walking country. Its hiking highlight, and one of Spain's finest walks, is the 12km (7.5-mile) Ruta de Cares, which winds through the vertiginous 'Divine Gorge' with steep drops, soaring cliffs and chilly tunnels to contemplate en route.

☛ SEE IT! *Fly into Bilbao and drive three hours west. The park's main access towns are Cangas de Onís, Arenas de Cabrales and Potes.*

© Olga Kashubin / Shutterstock

Breathe deep in Freycinet National Park & Wineglass Bay

AUSTRALIA // You might think this bay, a silvery crescent of sand hugging a shimmering lagoon, is named for its shape, or the fact that it's an excellent place to lounge with a glass of vino. But Wineglass Bay owes its title to something far less pleasant. In the early 1800s this corner of eastern Tasmania's Freycinet Peninsula was home to a whaling station, and the blood of the slaughtered whales turned the water as red as Cabernet. Today the exquisite bay is part of Freycinet National Park, a land of pink granite mountains, fragrant eucalyptus forests and sunsets like melted gold. Kayak in the tropical-coloured (but cold!) water, scramble over lichen-covered rock formations and watch white-bellied sea eagles swoop over feldspar cliffs.

☛ SEE IT! *Wineglass Bay is a 2½-hour drive from Hobart. Buses to the access town of Coles Bay take three hours.*

Plumb the depths of Siberia's astonishing abyss at Lake Baikal

RUSSIA // Ancient Baikal has stirred and sustained nomadic tribes, revolutionaries, artists and adventurers for centuries. Plummeting 1642m (5387ft) from its icy waves to its almost unfathomable bed, it's the deepest lake on the planet. Whether you swim in it, sail on it, swallow it (the waters are said to have magical properties), or just admire it from its 2000km (1240 miles) of shoreline, you'll find 'the Sacred Sea' to be nature at her most astounding.

☛ SEE IT! ***Visiting is easiest from Listvyanka village via Irkutsk; Severobaikalsk (on the BAM railway) is best for accessing trails.***

Admire Rome's ambition among Baalbek's ruins

LEBANON // Who knows why the Romans decided to build their most audacious temples in a backwater of the empire, but today's visitors to the 'City of the Sun' thank them for the effort. Baalbek's temples eclipse, in sheer scale, any other building projects the Romans ever attempted. Be reduced to ant-size amid the Temple of Jupiter's towering granite columns, then sit on an oversized limestone slab in the Temple of Bacchus' peristyle to take in this monument fit for giants.

☛ SEE IT! ***Baalbek is 90km (55 miles) from Beirut. The Bekaa Valley security situation changes rapidly. Check your government's latest travel advisory before visiting.***

Climb to monastic heights at Lhasa's Potala Palace

CHINA // Potala Palace is the largest monastery in the Buddhist world and one of the most recognisable buildings on the planet. Spread over 13 floors, with whitewashed walls cascading down the face of Marpo Ri (Red Hill), the palace was Lhasa's spiritual heart. Inside, the intricate, colourful murals and golden statues convey a sense of a vanished era but today more visitors than monks wander its passageways and painted chapels and the palace has the air of a memorial.

☛ SEE IT! ***The Potala looms over central Lhasa; make sure you're acclimatised to the altitude before tackling the steep walk to the top.***

Float beside bus-sized bergs in Ilulissat Kangerlua

GREENLAND // Welcome to the Greenland of your travel fantasies. This astonishing 40km (25-mile) ice-fjord, packed with bergs the size of apartment blocks, is fed by Sermeq Kujalleq, the world's fastest-moving glacier, which shifts at an average speed of 40m (130ft) daily. It's mesmerising to gaze upon monster icebergs and hear the almighty thunderclap when they fissure in the warmth of the summer sun. Watch the surreal treadmill from one of the boat tours plying the bay.

☛ SEE IT! ***Greenland's third-largest town, Ilulissat, is a spectacular spot to base yourself; drink in views of the ice-fjord from the bars and restaurants of its pretty rainbow-hued buildings.***

Whoop it up at the races in Siena's Piazza del Campo

ITALY // Ringed by Gothic palazzi, the great red-brick piazza has been Siena's social centre since the 12th century when the Council of Nine devised its pie-piece paving, frescoed its Palazzo Pubblico and constructed the 332-step Torre del Mangia. From its heights, you can look down on people picnicking on the warm paving stones – until it's time for the Palio, a spectacular bareback horse race dating from the Middle Ages. On race days (2 July and 16 August) find a good spot in the centre of the Campo and ride the wave of excitement, as competing city *contrade* (districts) swirl round the course tossing flags to banging drums. The race, when it comes, is a blur of hooves, dust, vibrant jockey silks and uproarious cheers. It lasts just 90 seconds, but it is electrifying.

☛ SEE IT! ***Siena is $1\frac{1}{2}$ hours from Florence by train. Piazza cafes sell seats for the Palio via the Tourist Office, or stand for free in the Campo.***

© mrohana / Getty Images

99

Go arty island-hopping in Naoshima

JAPAN // Naoshima is one of Japan's great success stories: a rural island that was on the verge of becoming a ghost town transformed into a world-class centre for contemporary art. Packed into this tiny spot in Japan's Inland Sea is a suite of architectural standouts, including museums and a boutique hotel designed by Tadao Ando. Traditional buildings have been turned over to creative installations, and dotted around the coast are several outdoor works, including the joyful *Pumpkin* by Yayoi Kusama. Art even fills the public bathhouse, where you can soak beneath a life-size elephant sculpture. Other islands in the Inland Sea have caught the avant-garde bug, too. When you've finished in the galleries of Naoshima, why not hop on a ferry to Teshima to register your heartbeat at Les Archives du Cœur?

☛ SEE IT! ***Ferries arrive from Uno (near Okayama) and Takamatsu (on Shikoku). Visit during the Setouchi Triennale for lots of activities.***

Yayoi Kusama's "Pumpkin". Photo © Naoharu Obayashi

100–199

Tilla-Kari (Gold-Covered) Medressa, was completed in 1660 and is the most ornate of Samarkand's great medressas, inside and out

© eFesenko / Shutterstock

100

Find a golden wonder at the Registan, Samarkand's exquisite Silk Road square

UZBEKISTAN // Even in Islam's competitive repertoire of important-looking mosques and glittering medressas (Islamic schools), Samarkand's Registan square stands out. It's an ensemble of majestic, tilting medressas and mosques – a near overload of majolica, azure mosaics and vast, well-proportioned spaces – that make up the most awesome sight in Central Asia. These beleaguered treasures have been battered by time and earthquakes, but their incredible craftwork and restoration under Soviet rule have kept them standing. Lovers of symmetry will be bowled over by the exquisite edifices flanking three sides of the square, which in medieval times would have been wall-to-wall bazaar. Ulugbek Medressa, on the west, was the first finished in 1420. Opposite is *Sher Dor* (Lion) Medressa, finished in 1636 and decorated with roaring felines. In between is the *Tilla-Kari* (Gold-Covered) Medressa, completed in 1660 with a mosque decorated with gold, to symbolise Samarkand's wealth at the time it was built.

SEE IT! ***Samarkand city was a key post on the Silk Road. Try to visit around sunset, when the tiles glow in the blue hour, and stay until the evening lights switch on to make everything sparkle.***

© Evgeniy Agarkov / Shutterstock

101

Top: Kirstenbosch contains over 7000 plant species. Bottom: the garden's Tree Canopy Walkway curves through the treetops

Inhale the perfume of a thousand scents in Kirstenbosch National Botanical Garden

SOUTH AFRICA // Kirstenbosch is one of the world's greatest botanical gardens, with its 5-sq-km (1.9-sq-mile) showcase of the Unesco-protected Cape Floral Kingdom, which has the planet's highest concentration of plant species. Wandering Kirstenbosch's garden paths and forest trails, you can see over 7000 floral species, including South Africa's national flower, the king protea, and other members of the heathery fynbos family. However, impressive as its fragrant plants walk and its valley of cycads are, Kirstenbosch is more than a garden.

To start with, there's its location on the side of Table Mountain, which the Skeleton Gorge and Nursery Ravine trails climb from the garden. The rocky slopes provide an epic backdrop for the Kirstenbosch Summer Sunset Concerts, held on Sundays from November to April, when Capetonians claim a patch of lawn for their picnic and chilled local bubbly.

Then there's the range of activities, including walking the 130m-long (427ft) Tree Canopy Walkway, which curves through the arboretum treetops and – for kids – splashing in streams and climbing wild almond trees.

☛ SEE IT! ***Kirstenbosch is 10km (6 miles) south of central Cape Town. The City Sightseeing hop-on hop-off buses stop here.***

© Quality Master / Shutterstock

© EcoPic / Getty Images

102

Stride grandly over hexagonal rocks at Giant's Causeway

NORTHERN IRELAND // Perched on a hexagonal basalt column, watching waves wallop Northern Ireland's Antrim coast, it's easy to imagine hulking Fionn MacCumhaill (aka Finn McCool) building this causeway across the sea to challenge Scottish giant Benandonner to a punch-up. The Celtic colossuses never did fight – Benandonner chickened out and destroyed most of the causeway while fleeing – but whatever the true origin of the rocks, the Causeway Coast is an extraordinarily evocative spot, with exceptional walking.

SEE IT! *The Causeway is 95km (60 miles) northwest of Belfast. June to August can be very crowded; try to visit early morning or after 4pm.*

103

Marvel at the Sheikh Zayed Grand Mosque

UNITED ARAB EMIRATES // Mosque design with full jazz hands; the Sheikh Zayed Grand Mosque (SZGM) does nothing by half. You want bejewelled? The marble courtyard is covered with a sprawling floral mosaic of semi-precious stones. Craftwork? The prayer hall holds the world's largest handmade carpet, fashioned from 2268 million knots. All that detailing is tied together by an architectural approach that includes nods to Arab, Persian and Moorish design, with columned arcades, reflective pools and a roof topped with 82 domes.

SEE IT! *The SZGM is in Abu Dhabi. Come twice: at 9am to beat the tour buses and in the evening to see the mosque lights.*

104

Meet pilgrims amid Amritsar's Golden Temple domes

INDIA // The main 16th-century Harmandir Sahib (Temple of God) at Amritsar's Golden Temple is a vision of perfection, rising from a shimmering pool, just part of a far greater *gurdwara* (religious complex) that comprises the Punjab's spiritual home. Sikh pilgrims who come here in their thousands are always keen to discuss their religion with visitors – these conversations are best enjoyed among shady colonnades, while chants echo gently along the marble surfaces of the ethereally beautiful building.

SEE IT! *Amritsar is well connected to Delhi. Everyone, regardless of faith, is invited to have a simple meal in the langar (dining room).*

© rarrarorro / Shutterstock

The National Museum of African American History & Culture's form was inspired by a Yoruban crown

105

Hear untold stories at the National Museum of African American History & Culture

USA // To comprehend the diverse African American experience and how it helped shape the USA, there is no better museum than this newest one in the Smithsonian's array. Just entering the bronze-tiered building, its design inspired by a Yoruban crown, feels remarkable. Exhibits spread over six main floors. The story begins in the underground galleries with early accounts of slavery and segregation. Walk around and there's a timbered slave cabin from a plantation, Harriet Tubman's hymn book and a shotgun shell from the 1963 bombing of the 16th Street Baptist Church in Alabama. Emmett Till's casket sits in a room of its own, with a solemn procession of visitors lined up to pay their respects. Move onward and Malcolm X's Koran, James Brown's black cape and President Barack Obama's inauguration invitation get their due. Interactive displays let you drive a car following the Green Book (a guide for African American road-trippers) and learn how to step dance. Mega queues attest to the museum's significance.

SEE IT! ***The museum sits on the National Mall in Washington, DC; Metro trains and Circulator buses stop nearby. Get a timed-entry pass online starting at 6.30am the day of your visit.***

106

Walk tall among the soaring stone columns of Karnak

EGYPT // Karnak is architecture on steroids. The Valley of the Kings may have thrashed it on this list but, for most Luxor visitors, Karnak gets equal billing. To the ancient Egyptians, this enormous religious complex of temples and sanctuaries, decorated by soaring obelisks, towering statues and mammoth pylons, was the earthly home of the god Amun-Ra. A roll call of Pharaohs from the New Kingdom's 18th to 20th dynasties have left their mark. These days the hypostyle hall in Karnak's Amun Temple Enclosure, with its forest of 134 papyrus-shaped columns, still stops people in their tracks as they contemplate its immense scale. Arrive early (it opens at 6am) to get a good couple of hours exploring before the 21st century intrudes in the form of tour buses.

SEE IT! *Karnak is 3.5km (2 miles) north of central Luxor. Sound and Light shows take place every evening.*

© Peter Seaward / Lonely Planet

107

Scale Matterhorn, a one-of-a-kind pyramid of a peak

SWITZERLAND // Rising dark and brooding like a shark's fin, the 4478m (14,692ft) Matterhorn has enthralled and infuriated hardcore mountaineers since 1865. The minute you clap eyes on this whopping great Toblerone of a peak, you'll become obsessed with capturing it on camera...or summiting it.

You'll need climbing experience and a guide to tackle its fearsome walls of rock and ice. Otherwise, a trip to the glitzy alpine resort of Zermatt is an easy way to get up close, either hiking or skiing in its shadow. Visit the Matterhorn Museum for an insight into the 19th-century mountaineers who pioneered the climb and the plucky souls who didn't make it because of rope-breaking tragedy.

SEE IT! *Matterhorn towers above the car-free town of Zermatt in the Swiss Alps, which can be reached by taking the train from Visp.*

© Olga Danylenko / Shutterstock

© Pawika Tongtavee / Shutterstock

© Michael Heffernan / Lonely Planet

108

Take a scenic road trip along dramatic Karakoram Highway

PAKISTAN // Steeper, more jagged and more densely packed than its Himalayan rival, northern Pakistan's Karakoram range gets our vote for the world's most dramatic, with over 20 peaks scraping the clouds over 7500m (24,600ft; including the world's second-highest peak, K2). Cutting right through the heart of these mountains is the astonishing KKH, an incredible feat of engineering that links the Pakistani capital at Islamabad with the Uyghur city of Kashgar in China's far-western Xinjiang region. The 1300km (800-mile) drive takes you through the sublime valley of Hunza and the witch-hat peaks of Gulmit, then over the 4714m (15,465ft) Khunjerab Pass to the Tajik fortress town of Tashkurgan, before passing high-altitude views of muscular Pamiri peaks to arrive in time for Kashgar's impressive Sunday market.

SEE IT! *Check the security situation in both Pakistan and China's Xinjiang region, which is currently in a state of political clampdown.*

109

Let Kruger National Park's animal magic cast a spell on you

SOUTH AFRICA // What makes Kruger stand out from other African game reserves? Well, the list is as long as the queues of thirsty wildlife at its waterholes, but the clincher is Kruger's accessibility. Where else can you drive your hire car right in through the front gate, strike out on a bush walk with a gun-toting ranger and then trundle off on a mountain bike all in the same trip? Numbers and variety of wildlife, staggering size and range of activities have made Kruger arguably one of the greatest places on earth to watch wildlife. Get ready for action-packed sightings of animals from the Big Five down. One morning, you may spot lions feasting on a kill, and the next, a newborn impala struggling to take its first steps. A trip here will sear itself into your mind.

SEE IT! *From airports at Nelspruit (Mbombela), Phalaborwa, Hoedspruit and Skukuza. Viewing is best from June to September.*

Hanli Prinsloo's Top Five Places

Hanli Prinsloo is a South African freediver, speaker and ocean conservationist. She is the founder of I Am Water, an ocean conservation trust dedicated to conserving and protecting the world's oceans.

BABYLONSTOREN WINE FARM & GARDENS, SOUTH AFRICA – This Cape Winelands estate offers orchards and beautifully landscaped fruit and vegetable gardens. The wine is great too.

CAPE KELP FORESTS, SOUTH AFRICA – Sinking into the swaying forests surrounding the Cape Peninsula is like stepping into a fairy-tale world of filtered light, blue urchin gardens and scattered sea stars.

SWEDISH WEST COAST – Spending most of my twenties on Sweden's west coast, I fell irrevocably in love with these rugged islands. The Swedish concept of *livsnjutare* (one who loves life) is more tangible here than anywhere.

BAJA EAST CAPE, MEXICO – Where the cactus meets the blue Sea of Cortez is a playground for ocean lovers. Whether you're curling off a rocky point or diving down into what Jacques Cousteau called 'the world's aquarium', the old-time feel is a favourite escape.

KALK BAY VILLAGE, SOUTH AFRICA – This small artisanal fishing village with crooked cobblestone roads and myriad bakeries and coffee shops is nestled between the rich ocean and majestic mountains of the Table Mountain National Park.

© Peter Marshall

Greet a geisha in the laneways of Kyoto's Gion District

© cowardlion / Shutterstock

JAPAN // With their elaborate kimonos, theatrical make-up and immaculate up-dos, the mysterious geisha usually confine their appearances to special occasions and best-selling novels. Yet if you're lucky you may catch a glimpse of them shuffling by as they go about their daily business in Gion. Kyoto's largest entertainment district by the mid-18th century, Gion is still a stronghold of Japanese geisha culture. Despite thronging tourists, it maintains an air of tradition, particularly in the little back-lanes. Behind the closed doors and shuttered windows of 17th-century teahouses, kaiseki restaurants and exclusive bars await, signalled by glowing lanterns.

SEE IT! ***See geisha at a scheduled event, like the* odori *(dance) performances.***

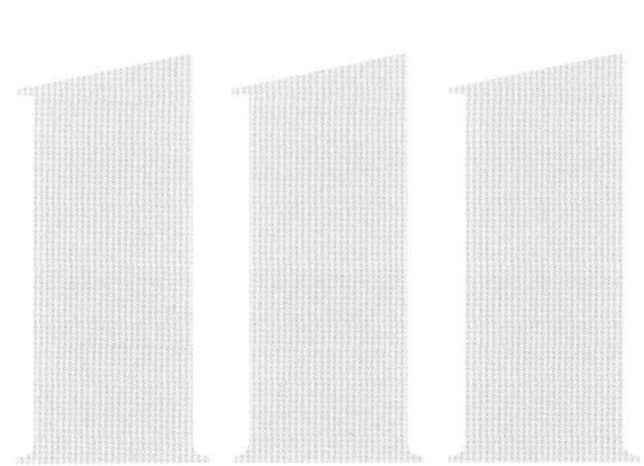

The Quebrada de Humahuaca's vivid, alien-looking canyons were sculpted by rivers

See a riotous rainbow of colours in the landscapes of the Quebrada de Humahuaca

ARGENTINA // Prepare to be dazzled. Stretching north through Jujuy province towards the border with Bolivia, the Quebrada de Humahuaca is a spectacular valley of dry, river-scoured canyons overlooked by mountains striped with a spectrum of colourful layers of sediment. The light here reveals a landscape in technicolour, as though the dimmer switch for the sun has been turned to maximum, illuminating every shade of creamy white, moss green, pale pink and rusty red in the canyons and mountains. Nowhere is the ever-changing palette more vivid than at Purmamarca's Cerro de los Siete Colores (Hill of Seven Colours), which positively glows in the early morning sun and evening light. Near the town of Humahuaca, the mountains of the Serranía de Hornocal form a jagged row of teeth-like rocks in a kaleidoscope of burnt oranges, saffron yellows and burgundies.

Dotting the valley are dusty indigenous villages, with cobbled streets, pretty adobe churches and market stalls selling ponchos and textiles in colours that echo the landscape. Traditional Andean culture thrives: tempting smells of home-cooked *locro* (stew) and the strumming of folk music carry on the breeze.

SEE IT! ***Reach the Quebrada by road from Jujuy city. February's carnival is worth seeing but days are hot (nights are freezing).***

Explore ingenious engineering in Florence's Duomo

ITALY // Get yourself on the outside of some biscotti and a stiff coffee before ascending the 463 steps into the cupola that crowns Florence's magnificent Duomo. The going is steep, but how often do you find yourself nose-to-nose with Renaissance frescoes the calibre of Giorgio Vasari's *Last Judgment*? With each step, the marvel of Brunelleschi's octagonal dome unfolds, the stairway squeezing between an inner and outer dome. The reward of reaching the summit: an unforgettable 360-degree panorama of one of Europe's most beautiful cities.

☛ SEE IT! ***To visit the cupola make a reservation online (at least a month ahead in high season) or at the ticket office's Ticketpoint.***

Get a taste of monastic life atop peaceful Kōya-san

JAPAN // Nestled in northern Wakayama Prefecture, Kōya-san is as much about the journey as the destination. The train trip winds through valleys and mountains before the final cable-car leg up to the peaceful setting of thickly forested Kōya-san itself. Founded in the 9th century, more than 110 temples make up this monastic complex, the headquarters of the Shingon school of Esoteric Buddhism. Stay overnight in temple lodgings for an insight into the life of a Japanese monk, complete with morning meditations and vegetarian Buddhist cuisine.

☛ SEE IT! ***From Osaka's Namba station, take the Nankai Railway to Gokurakubashi then the cable car up the mountain.***

Take an epic road trip around the Ring of Kerry

IRELAND // Winding past pristine beaches, medieval ruins, seaside villages, scattered islands, moody mountains and shimmering loughs, this famous loop around the Iveragh Peninsula tempts travellers to take unplanned detours and leisurely stops. Spinning off the main route, the scenic Skellig Ring offers yet more adventures along roads too narrow for tour buses. Silent movie star Charlie Chaplin was so smitten that he spent annual family holidays in the village of Waterville, where his bronze statue stands on the seafront.

☛ SEE IT! ***Most people drive the 179km (111-mile) loop. The best time to cycle the route is May or June; roads are busiest in July and August.***

Beach-hop on Brazil's Fernando de Noronha

BRAZIL // In one of the most beach-blessed countries on Earth, this volcanic archipelago still stands out. Flung more than 350km (220 miles) off the coast, its 21 islands are a playground of parakeet-green hills, diamond-bright sands and waters aswim with sea turtles and dolphins. With water too warm to require a wetsuit, it's a diver's playland. Everyone else can spend their time toasting on the beaches and hiking amid the crumbling Portuguese forts.

☛ SEE IT! ***Flights from Recife and Natal take about an hour. Once here, ride the bus or rent a buggy.***

© Simon Crockett / Getty Images

West Bay Cliffs, part of the Jurassic Coast, near Burton Bradstock in Dorset

116

Find spits, stacks and ancient beasts on the Jurassic Coast

ENGLAND // First came the skull, a strange, slender thing, over 1m long, found by Joseph Anning in 1811. Then his sister Mary found the rest: the skeleton of a great creature from the sea. It was the first complete ichthyosaur to be found, and it turned our understanding of the past on its head.

Discoveries – of ichthyosaurs, dinosaurs and more – kept coming. This is the Jurassic Coast, stretching from Swanage in Dorset to Exmouth in Devon, its unique jumble of clay, limestone and sandstone covering 185 million years of history in 150km (95 miles). Its geology makes this World Heritage site a particularly fine place to uncover fossils – and one of Britain's most beautiful spots. At Durdle Door a grand arch dips into the sea, Chesil Beach is a vast stretch of shingle, Lyme Regis has steep, fossil-packed cliffs, and Burton Bradstock gorgeous sands backed by rock.

This land of spits, stacks, beaches and frothy ale is ripe for adventure, and stars in everything from smugglers' yarns to Jane Austen. You can swim, kayak, surf or hike coastal trails – or make like the Annings and fossick for glimpses of long-lost worlds.

☛ SEE IT! *Lyme Regis, Weymouth, Exeter and Swanage are good bases. The Jurassic Coaster bus connects towns and villages.*

The remote Westfjords are home to fishing villages, seabirds, hiking trails and great stretches of wild, windswept coastline

© ronnybas / Shutterstock

117

Blow your mind with the Westfjords' big skies, rainbows, ridges, sorcerers and lots of seabirds

ICELAND // The Westfjords is Iceland's magnificence, hidden in plain sight. Only about 10% of visitors ever come to this remote region, and so the gorgeous series of rainbow-topped fjords remains a fresh, untrammelled zone. It's crowned in the north by the hiking reserve Hornstrandir, an enclave of Arctic foxes and seabirds. The Strandir region is known as the former home of Iceland's sorcerers, and it remains a fog-shrouded, end-of-the-road kind of place, with just a few village outposts and razor-backed ridges in between. As you head south through the undulating fjords, you'll encounter gigantic waterfalls, like glorious Dynjandi, which really roars as it catches all of the runoff of nearby mountain valleys.

In the south, the white- or red-sand beaches are well worth the trek. One reward is the Latrabjarg bird cliffs, where puffins, razorbills, guillemots, cormorants, fulmars, gulls and kittiwakes nest from June to mid-August. Small fishing villages dot the region, offering a respite from the wilds with a warm bowl of soup and hot, fresh bread. Not to mention a good story or two.

SEE IT! *Plan for plenty of time: unpaved roads wind around secluded fjords and over rugged mountain passes. Especially outside of summer, it's essential to check road conditions at www.road.is.*

© Alessandro Colle / Shutterstock

© Monica Hernandez Ahman / Shutterstock

118

Climb aboard for glacier-capped Jungfraujoch

SWITZERLAND // If you haven't heard of Jungfraujoch, you need to get on board. There's a reason why more than a million people a year venture up 3454m (11,332ft) to Europe's highest train station. The icy wilderness of deeply crevassed glaciers and 4000m (13,120ft) turrets that unfolds at the top is out of this world.

Within the sci-fi Sphinx meteorological station, you'll find viewpoints and an Ice Palace gallery of other-worldly sculptures. The phenomenal journey is the icing on the cake: the last stage of the cogwheel train ride burrows an audacious route right through the heart of the Eiger, delving into a tunnel that took 3000 men 16 years to drill, back in 1912.

SEE IT! *Jungfraujoch is in the Swiss Alps. From the Bernese Oberland resort town of Interlaken, the rail journey is 2½ hours each way. Go first thing in the morning to give the crowds a slip.*

119

Relish a Caribbean colourfest in Cartagena's Old Town

COLOMBIA // Having pierced the rind of *Las Murallas* (the city walls), an explosion of tangy citrus colours confronts you in Cartagena's Old Town, on the Caribbean coast of Colombia. Cartagena's sturdy walls are a direct legacy of the visit, in 1586, of one Francis Drake, who invaded the Colombian city on behalf of Elizabeth I and stripped it of pretty much everything he could make off with. In the aftermath of the attack and occupation, the city concluded that it needed better defences. Once you're in the walled town, the best way to explore the historical districts of El Centro and San Diego is simply to meander, tasting local treats and soaking up the lively street life.

SEE IT! *Cartagena is 1000km (620 miles) north of Bogotá – roughly an hour's flight. It's warm year-round, and December to March is dry season.*

© Darshana Abraham / 500px

Top: Adam's Peak is scaled by pilgrims and tourists, many of whom aim to summit just before dawn breaks. Bottom: Sunrise at the summit

120

Hike through the night to watch dawn break over sacred Adam's Peak

SRI LANKA // Pilgrims have been climbing Sri Pada (Adam's Peak) for centuries. Christians come to see the spot where Adam allegedly first set foot on earth after being cast out of paradise; Buddhists come to see the huge footprint of Buddha that crowns the peak. As you set off in the pre-dawn chill you'll follow a dotted line of pilgrim's torches bobbing uphill through the darkness. Time your walk well and you'll arrive on the summit just in time to catch the first blush of dawn crashing across the waves of ridges surrounding the holy mountain. This is communal hiking at its most jovial.

SEE IT! ***Most pilgrims start their hike from Dalhousie, 140km (87 miles) from Colombo. Come on a poya (full moon) day for maximum pilgrim colour, but be prepared for long queues up the mountain.***

© Melinda Nagy / Shutterstock

© Martin Capek / Shutterstock

© Nowak Lukasz / Shutterstock

121

Emulate an elk on the soaring heights of Denali

USA // Although Mt Everest is the world's highest mountain by altitude, Alaska's Denali (6190m/20,310ft) is actually 1800m (6000ft) taller because it starts from a much lower base. The Athabascans christened it the 'Great One', and few who have seen this fearsome bulk of ice and granite would disagree. Observed from the road in its eponymous national park, Denali chews up the skyline, dominating an already stunning landscape of tundra fields and polychromatic ridgelines. If you're lucky, the odd moose or bear will pop into the picture as you contemplate the scene.

The mountain inspires a take-no-prisoners kind of awe, especially among climbers, more than 1000 of whom arrive each year hoping to reach the horizon-bending view of a lifetime. Mid-June to early September is the park's main season.

☛ SEE IT! *Denali is a five-hour ride north of Anchorage and two hours south of Fairbanks. The most fun approach is on the Alaska Railroad.*

122

Drive the roof of the world on the Pamir Highway

TAJIKISTAN // One of the world's great road trips, driving the 'roof of the world' along the Pamir Highway through Tajikistan is a rough-and-tumble experience through one of the last truly remote parts of the planet. Built by the Soviets in the 1930s, the highway was the only thoroughfare into Tajikistan. For the most part, the route is an unpaved, rutted, one-lane road through stark, arid mountains at impossibly high altitudes.

Although officially the Pamir Highway refers to a section of the M41 between Khorog in Tajikistan and Osh in Kyrgyzstan, unofficially the route starts in the Tajik capital Dushanbe, and some travellers opt for a detour following a smaller road along the Pyanj River and Afghanistan border in the Wakhan Corridor. Along the way, you stop in timeless Pamiri villages, pass by lost hilltop forts and drink countless cups of tea.

☛ SEE IT! *Pre-planning is a must: you need to hire a 4WD and an experienced driver to make this journey.*

© chrisadam / Getty Images

© Matteo Colombo / Getty Images

123

Watch the ocean's great beasts rise up in Kaikoura

NEW ZEALAND // The waters off this postcard-pretty South Island peninsula are an all-you-can-eat buffet for whales, thanks to a current-driven upswelling of deep-sea nutrients. Don your rain jacket and climb aboard a whale-watching vessel to spot – season depending – migrating humpbacks, breaching right whales, bus-sized blue whales, and sleek, grinning orcas. But the big draw is the mighty sperm whale, its bulbous body emerging from the water with surprising grace, immense tail slapping the surface into a froth. They're here year-round, and tours promise 80% chances of sightings. If you're quick with your camera, you'll catch unbeatable shots of whale tails framed against the snowy Kaikōura Range. Also look out for pods of playful dolphins, enormous albatrosses, and piles of sleepy fur seals lazing on the rocky shore.

SEE IT! *Kaikoura is 2½ hours north of Christchurch or two hours south of Blenheim by car. There are buses from Christchurch too.*

124

Beat the crowds at Chichén Itzá

MEXICO // In the world of over-trammelled tourist sites, Chichén Itzá falls into the 'busy, but worth it' category. Mexico's greatest Mayan ruins hail from the post-classic period of Mesomerican history and juxtapose grandiose pyramids with more mysterious astronomical phenomena, such as the ancient on-site observatory known as 'El Caracol'.

Since being inscribed as one of the seven 'new' wonders of the world in 2001, tourism at Chichén has increased significantly, with busloads of visitors invading the site daily. But, arrive early in the day and blot out the crowds, and you'll quickly start to wrestle with that rising thrill of discovery. Chichén's muscular stone temples, great ball-court and broad, chunky platforms are the Gothic cathedrals of the Mayan lowlands – built to last.

SEE IT! *Chichén Itzá is two hours from Cancún on the Yucatán Peninsula. For the best experience, arrive soon after 8am.*

Opposite, from top: boardwalks crisscross Plitvice Lakes National Park; Versailles was the main home of the French royal family for over 100 years

125

Be absorbed by a turquoise water world at Plitvice Lakes National Park

CROATIA // A striking ribbon of crystal water and gushing waterfalls in the forested heart of continental Croatia, Plitvice Lakes National Park is excruciatingly scenic and has to be one of the most singular parks in the world. There are 16 lakes – from 4km-long (2.5-mile) Kozjak to reed-fringed ponds – all in dazzling shades of turquoise that are a product of the karst terrain. Travertine expanses covered with mossy plants divide the lakes, while boardwalks allow you to step right over this exquisite water world and trails lead deep into beech, spruce, fir and pine trees. Bears, wolves and deer roam here, but perhaps you're more likely to catch a glimpse of a swooping hawk or the occasional black stork. Hoof it or explore by free park boats and buses; definitely don't miss the boat trip to the base of 78m (255ft) Veliki Slap, the tallest waterfall in Croatia. It's surreal to think that this cool blue World Heritage-listed site was blighted by bloody conflict during the war of the 1990s.

SEE IT! ***Plitvice is located in Croatia's Adriatic hinterland, a couple of hours from the capital Zagreb. Avoid often-overcrowded July and August.***

Roam the regal corridors of France's stateliest château, Versailles

FRANCE // You can try to imagine Versailles simply as a place where people ate, drank, worked and slept. But seriously, how many houses in the world have 700 rooms, 2153 windows, 67 staircases, 800 hectares of garden, 2100 statues and sculptures, and enough paintings to pave an 11km (7-mile) road? Château de Versailles is one almighty crash pad.

Even more striking than its size is the ostentatious opulence. French 'Sun King' Louis XIV transformed his father's hunting lodge southwest of Paris into a monumental palace to house his 6000 sycophantic courtiers in the 17th century. It was the kingdom's political capital and seat of the royal court from 1682 until the French Revolution in 1789.

Today it's a glittering evocation of French royal history and the conspiring, romancing and backstabbing drama that went on behind its doors. Gaze at your reflection in the shimmering Hall of Mirrors, see horses prance in the stables, and watch fountains dance to baroque music in summer in the immaculately landscaped formal gardens.

SEE IT! ***Take the RER C5 from Paris' Left Bank RER stations. Avoid Monday (shut), Tuesday and Sunday (ultra-busy). Pre-purchase tickets online.***

125

126

Top: despite the ravages of fire, war and shifting desert sands, Persepolis' Gate of All Nations endures. Bottom: two-headed griffin sculpture

© Leonid Andronov / Shutterstock

127

Trace the march of history through the friezes of Persia's Persepolis

IRAN // Few ancient ruined cities possess as much power to beguile as Persepolis. This Unesco-listed site was the epicentre of the great Persian empire of antiquity, and the monumental staircases and imposing gateways leave you in no doubt how grand this civilisation was, just as the broken and fallen columns attest that its end – at the hands of Alexander the Great – was merciless. Like the ancient foreign dignitaries were, visitors are ushered in through the immense Gate of All Nations, still guarded by two colossal winged bulls. Thousands of bas-relief stone soldiers line the walls of the Apadana Staircase, and elaborate panels show delegations from across the Achaemenid Empire bringing local gifts to the king. In something of a roll call of history, the Ethiopians begin the frieze and are joined by Arabs, Thracians, Kasmiris, Parthians, Cappadocians, Elamites, Egyptians and Medians. The Palace of 100 Columns was one of the two main reception areas for meetings with the king, and enough of the poignant ruins and broken columns remain today to let visitors imagine the height of its grandeur.

SEE IT! ***Shiraz is the most convenient base for Persepolis – rent a taxi, or tour for the day. Arrive early to beat both crowds and heat.***

© Anna Azimi / Shutterstock

128

Scale Sigiriya, Sri Lanka's top rock star

SRI LANKA // Rising sheer from surrounding plains, the grand outcrop of *Sigiriya* (Lion Rock) is perhaps Sri Lanka's most dramatic sight. Clamber up gravity-defying staircases past the huge stone lion paws that mark the old palace gateway, and the near-vertical rock walls eventually yield to a flat summit containing the ruins of an ancient civilisation, as well as spellbinding vistas. The surrounding landscape – lily-pad-covered moats, water gardens and quiet shrines – only adds to Sigiriya's charm.

SEE IT! ***Sigiriya is east of Inamaluwa, between Dambulla and Habarana; buses run to the gardens. It's 1200 steps to the top.***

129

Savour the medieval charm of Bruges at its midweek Markt

BELGIUM // The exquisite city of Bruges has one of Europe's best-preserved medieval cores. Laced by canals and cobbled lanes, it centres on the splendid market square, the Markt, garlanded by guildhalls and cafes. First hosting markets in AD 958, it still fills every Wednesday with stalls selling fresh produce and ready-to-eat treats.

Towering 83m (272ft) above the Markt is the city's most defining sight, the Belfort, Bruges' octagonal medieval belfry; braving its 366 steep, narrow steps rewards with an incredible panorama of the city.

SEE IT! ***Visit midweek or outside high summer (and December, when the Christmas market sets up) to avoid day-tripping crowds.***

130

Enjoy an *Arabian Nights* moment at Jaisalmer Fort

INDIA // The honey-coloured fortress of Jaisalmer rises from the desert plains of western Rajasthan like a mirage, encircled by 99 mighty bastions. Inside, the fortress is a crazed tangle of ornate rooms, graceful *havelis* (merchants' houses) and ancient Rajput temples that almost collapse under the weight of their carvings and statuary. To visit Jaisalmer is to step into a fantasy sandcastle seemingly carved out of the desert itself. Its fragile sandstone structures won't be around forever, but while they are, they're a truly magical sight.

SEE IT! ***The fort is the heart of Jaisalmer, but most people stay in surrounding bazaars rather than inside in the fragile havelis.***

© papi8888 / Shutterstock

131

Scuba dive into Mu Koh Similan's soft-coral paradise

THAILAND // Imagine encountering an underwater world straight out of *Finding Nemo*. Sunlight dapples amazingly clear turquoise water, illuminating banks of gently swaying soft corals around which swim dazzlingly coloured tropical fish. Flipper-kick your way through dramatic underwater gorges alongside schools of sharks and barracudas. No wonder the Mu Koh Similan National Park is Thailand's premier dive spot. Embracing an archipelago of 11 islands adrift in the Andaman Sea, this is a fragile world; coral bleaching has killed off many hard corals. The park is closed to the public from approximately mid-May to mid-October and overnight stays on the islands are also banned to help protect the ecosystem. It is possible to visit for the day, though, if you want to go hiking, snorkelling or simply work on your tan on the superb, palm-fringed beaches.

☛ SEE IT! *The only way to access the park is on boat trips – either day tours or multi-day liveaboards – from Phuket or Khao Lak.*

Will Bolsover's Top Five Places

Will Bolsover is the founder of Natural World Safaris. He has guided extensively throughout Africa and helped establish the first ever gorilla-tracking safaris in Gabon. With NWS he champions the efforts of some of the world's leading conservation charities, including Wild Aid and African Parks.

SVALBARD, NORWAY – Small ship exploration through the icescapes of Svalbard is hard to beat. Vast glaciers, the frenetic bird cliffs, walrus haul-outs and of course, the King of the Arctic (aka polar bears) are all highlights.

CONGO BASIN, REPUBLIC OF CONGO & CENTRAL AFRICAN REPUBLIC – Odzala-Kokoua National Park and Dzanga-Sangha National Park are home to some of the most unique wildlife on our planet.

MADAGASCAR – The fourth-largest island in the world is home to vast numbers of endemic fauna and flora that astound even the most avid wildlife fan: indri lemurs, leaf tailed geckos, pristine seas and dramatic rainforests.

KAMCHATKA, RUSSIA – The planet on steroids! Only accessible by sea or air, the peninsula has some of the biggest brown bears on earth. With thermal springs, active volcanoes and pure wilderness, it's the backdrop for one of the best travel adventures around.

ELLERMAN HOUSE, CAPE TOWN – A rare luxurious choice, as I tend to prefer rough and remote. But Ellerman House is a hand-on-heart recommendation for honeymooners looking for Cape Town's ultimate address – and a difficult hotel to leave!

© Sophy Roberts

© Jonathan Gregson / Lonely Planet

© Hung_Chung_Chih / Getty Images

132

Have an unguided adventure at Mana Pools National Park

ZIMBABWE // You've hardly driven through the gates of this 2200-sq-km (850-sq-mile) reserve before you spot a herd of elephants, just casually stomping by – nothing to see here! It's like that in Mana Pools National Park, a place so rich in wildlife you'll almost – almost – stop being amazed every time you lay eyes on a zebra or hyena. Herds of buffalo roam the savannah, hippos and crocodiles wallow in the pools, cheetahs and Cape wild dogs slake their thirst on the banks of the Zambezi, and rare birds flit through forests of ebony and wild fig. And almost unbelievably, you're allowed to walk through all of this without a guide. A canoe safari means drifting past baobabs and baby elephants – and dodging hippos.

SEE IT! *Arrive at the park's airstrip via chartered flight. The May to October dry season is the best time to visit.*

133

Join the army – China's Army of Terracotta Warriors

CHINA // When it's your time to go, go out in style is our motto. So hats off to Chinese emperor Qin Shi Huang, who constructed an army of terracotta warriors to accompany him into the afterlife. The discovery of the 8000-strong army of clay warriors was a happy accident; in 1974, farmers drilling a well uncovered an underground vault that eventually yielded the terracotta soldiers and hundreds of chariots and horses.

The site is so famous that you probably know many of its most remarkable attributes – like how no two soldiers' faces are alike, or how the first emperor still lies buried nearby in an unexcavated tomb, surrounded by a scale model of his empire, complete with rivers of mercury. The visiting crowds can resemble the army itself.

SEE IT! *Buses go to the Terracotta Warriors from Xi'an Railway Station via Huaqing Hot Springs and the Tomb of Qin Shi Huang.*

© Jonathan Gregson / Lonely Planet

© Dan Breckwoldt / Shutterstock

134

Splash about in the African sun on Lake Malawi

MALAWI // Kayaking to a desert island on a lake in the middle of Africa – not a bad way to spend an afternoon. A drowned section of the Rift Valley, Lake Malawi was 'discovered' by Livingstone, so thank the good doctor as you lounge in a hammock, munching fresh banana bread in Africa's friendliest country. This watery Serengeti is one of the world's best freshwater dive spots, with around 1000 mostly endemic fish species – more than any other inland body of water. Look out for dolphinfish, which swarm towards divers' torches on night dives, and the multicoloured members of Africa's largest fish family, cichlids. Their home is nicknamed the 'calendar lake' for being 365 miles long, 52 miles wide (587km by 84km) and fed by 12 rivers.

SEE IT! *There are resorts, backpacker lodges and campsites in easygoing beach towns such as Cape Maclear and Nkhata Bay. Near the former are eco-camps on Mumbo and Domwe Islands.*

135

Be dwarfed by the majestic Ramses II at Abu Simbel

EGYPT // You'll be served a taste of pharaonic ego when you stand, dwarfed, in front of the 20m-high (66ft) statues of King Ramses II that guard the entrance to Abu Simbel's Great Temple. Along with the neighbouring Temple of Hathor, this complex was carved out of the mountain on the west bank of the Nile between 1274 and 1244 BC and stood sentinel over the road south for centuries. Abu Simbel's temples were saved from destruction during the building of the Aswan High Dam. An international rescue campaign relocated them block by gargantuan block to their current site on the shore of Lake Nasser. Today Abu Simbel is as much a monument to this modern engineering triumph as it is to Ramses II's puffed-up pride.

SEE IT! *Abu Simbel is 288km (179 miles) south of Aswan. Get the temples to yourself by staying overnight and visiting in the early morning.*

© Ben Cranke / Getty Images

Antarctica's Ross Ice Shelf covers an area as large as Spain – but it's shrinking fast

136

Feel the frozen force of Antarctica's Ross Ice Shelf

ANTARCTICA // Could this be the end of the world? It sure looks like it. Covering some 472,000 sq km (182,000 sq miles), an area approximately the size of Spain, the colossal Ross Ice Shelf seems to stretch away to infinity. When Captain James Clark Ross discovered it in 1841, he called it the Victoria Barrier (after the British monarch). And 'barrier' is an appropriate description for this gigantic shelf – up close it resembles a natural fortress that practically screams 'get back!'.

In places, the ice is up to 1000m (3280ft) thick, and if you think that's impressive, consider that 90% of it is submerged below sea level. For such an enormous lump, it's surprisingly nimble and moves as fast as 1100m (3610ft) per year. It actually floats, and calves huge icebergs annually, most famously B-15, the largest iceberg on record, measuring 295km (180 miles) in length. As the ocean warms, the ice shelf is melting at an alarming rate – 10 times faster than initially predicted. See it while you can.

SEE IT! ***Most visitors reach Antarctica via expedition tours (look for environmentally conscious operators), many of which board in Ushuaia at the southern tip of Argentina.***

137

Get rhapsodic at Singapore's Gardens by the Bay

SINGAPORE // Whether you're a garden lover or not, it's impossible not to be knocked out by this fantasyland of space-age biodomes, whimsical sculptures and conservatories housing some 217,000 plants from 800 species. It includes a Flower Dome replicating the Mediterranean climate and a Cloud Forest with a waterfall, but it's hard to trump the Supertrees 18 soaring, steel-clad structures adorned with vegetation, some joined by a skyway with knockout city views. These vegetated pillars twinkle nightly during the Garden Rhapsody light-and-sound show.

SEE IT! *Accessed from Bayfront MRT/metro station (Exit B), the gardens are free but there's a fee for the conservatories and Supertrees.*

138

Revel in jazz, history and voodoo in the French Quarter

USA // The heart of New Orleans beats in the French Quarter, the city's 300-year-old core. Those postcard images you see of elegantly aged, pastel-coloured buildings, cast iron balconies and courtyard gardens? Here. Voodoo shops for buying chicken-feet talismans and mojo bags? Also here. So are storied jazz clubs. This is the city where the genre originated, after all. Add in the slew of buskers, art galleries, streetcars and neon-lit, 24-hour bars, and you've got a sense-seducing district beyond compare.

SEE IT! *The Canal, Rampart and Riverfront streetcars skirt the edges of the district. March to May are good for weather and music.*

Bob like a cork in the super-salty Dead Sea

JORDAN // Sitting between Jordan, Israel and the Palestinian Territories, the Dead Sea is the lowest point on the earth's surface at 431m (1412ft) below sea level. Entering its super-salty water feels more like wading into syrup. But kick your legs off the bottom and you're naturally buoyant and able to sit, roll over, and lift arms and legs in the air all while staying afloat. It's tremendous fun, and the visceral landscape – all glittering blue water, blinding-white salt deposits and swooping craggy mountains – is one of the world's most dramatic settings to take a dip.

SEE IT! *You can access the Dead Sea at beaches on the Jordanian side and from both the Israeli and Palestinian West Bank shores.*

137

© BigBoom / Shutterstock

© posztos / Shutterstock

Marrakesh's Djemaa El Fna is the city's main square, and a 'Masterpiece of World Heritage' according to Unesco

140

Roll up for the street circus on Marrakesh's Djemaa El Fna

MOROCCO // Chaotic and enchanting in equal measures, Djemaa El Fna is the vibrant heart of Marrakesh. This main square is a hub of hoopla, *halqa* (street theatre) and communal noshing that has been going since medieval times. During the day it's all snake-charmers, hawkers, henna tattoo artists and juice-sellers. Once the sun sets though, the square transforms into a raucous mash-up of music, mayhem and feasting that is part circus, part open-air concert.

Some say this was Marrakesh's public execution site in the 11th century – the square's name can be translated as 'assembly of the dead'. Another theory is that it's named after an earlier iteration of the Koutoubia Mosque – 'djemaa' can also mean mosque – that collapsed during the 1755 Lisbon earthquake, burying worshippers inside. From these macabre beginnings, the Djemaa has evolved into today's carnivalesque centre, which Unesco proclaimed a 'Masterpiece of World Heritage' in 2001. Weave through the crowds, past lantern-lit stalls offering grills, tajines and snail broth, to find sub-Saharan Gnaoua musicians hypnotising huddles of spectators and you'll understand why.

SEE IT! ***For the full Djemaa experience, arrive at sunset. Bring lots of tipping coins.***

141

Hike the Turkish coast on the Lycian Way

TURKEY // Turkey's Teke Peninsula, a gorgeous chunk of beach resorts and hill villages among the olive groves and pine forests, was the ancient stomping ground of the Lycians. These contemporaries of the Romans, who had their heyday around 100 BC, left ruined cities, cliff tombs and amphitheatres. Explore their bygone kingdom on the Lycian Way, a waymarked 540km (336-mile) footpath from Fethiye to Antalya, winding through the foothills of the Western Taurus Mountains and along the Mediterranean coast. There are magnificent ruins and whitewashed harbour towns to rest your feet.

SEE IT! *The closest airports are in Dalaman and Antalya. You can walk short sections of the trail, which takes a month in its entirety.*

142

Enter the abode of ancient gods at Luxor Temple

EGYPT // The columns raised by New Kingdom Pharaohs Amenhotep III and Ramses II stand in the heart of modern Luxor. This monument is a stone spectacle of reliefs, shrines and statues that shines a light on the power and wealth of Egypt's Pharaohs. Come in late afternoon to wander through vast chambers and courts as the stones take on a golden glow, then again after dark when the illuminated columns create an eerie spectacle.

SEE IT! *The temple is in downtown Luxor. The 9pm closing time means you can see a lit-up temple without having to shell out for a Sound and Light show.*

© Philip Lee Harvey / Lonely Planet

Photo © Katy Murenu. Murals © www.kaff-eine.com

143

Hold on tight to see the lofty rock-hewn churches of Tigray

ETHIOPIA // Tigray's churches reek of mystery and adventure, and several require visitors to have nerves of steel – to reach Abuna Yemata Guh you'll need to climb some vertical sections of rock using toeholds (yes, you'll need to take your shoes off...), then navigate a perilously narrow ledge over a 200m (656ft) drop. Carved into mountaintops and sheer cliffs, these remarkable churches were virtually unknown to the outside world until the mid-1960s. Some think these remote locations were used to avoid Muslim raiders, others believe the lofty perches diminish the distance to God. Either way, their confines are covered in frescoes that date back over a millennium. Outside is one hell of a view.

SEE IT! *Mekele is a good base, linked by air to Addis Ababa. If you're patient, exploration by public transport is possible. Otherwise hire a guide and 4WD. Visit October to February.*

144

Road trip along the Silo Art Trail

AUSTRALIA // Far from the opera houses and flat-white coffees of the coastal cities, Victoria's remote Wimmera-Mallee region stretches vast and dry, dotted with towns with names like Brim and Sheep Hills. Why trek out here? To see Australia's largest outdoor art gallery, of course. In recent years, this agricultural country's tall metal grain silos have been transformed by prominent artists into enormous murals showcasing local community members, much to the rapture of art fans worldwide. Drive the 200km (125-mile) trail from silo to silo, taking in portraits of Aboriginal residents, local farmers, student athletes and kelpie sheepdogs, all dozens of metres high. It's a brilliant blending of art and history, past and present, the practical and the whimsical.

SEE IT! *The original silos can be visited in a (long) single-day drive; there are now more than 30 silos and water towers in an expanded version of the trail.*

Top: trekkers on Mutnovsky, a great volcano with a geyser field at its base. Bottom: brown bears are found in Kamchatka in large numbers

© Markus Renner / Getty Images

145

Discover the primeval north on the Kamchatka Peninsula

RUSSIA // It takes preparation to visit Kamchatka, the 1250km (775-mile) peninsula jutting from Russia's far east, between the Sea of Okhotsk and the lashing North Pacific. And we're not talking first-aid-kit-extra-batteries preparation. We're talking guide-with-a-rifle, helicopter-drop-ins, avalanche-safety-gear preparation. Kamchatka is wilderness at its most wild, with the largest active volcano in the northern hemisphere, 15,000 brown bears, thundering reindeer herds and snowmelt-engorged rivers capable of sweeping away a tank.

Duck your head and climb aboard a helicopter to soar over the fiery eyes of volcanoes and the milky blue lakes of the Valley of Geysers. Then land for a soak beneath the midnight sun in a natural hot spring. Snowshoe into the lonely mountains, tent strapped to your back. Or ride a boat across the churning grey seas for the chance to spot an orca breaking the waves. You'll almost certainly see dolphins, seals, walruses and shrieking seabirds of all stripes along the jagged coastline. For anglers, the trout and salmon fishing is the stuff of whispered legends.

SEE IT! ***Visiting independently is possible, but a small tour is much easier. There are daily flights to Moscow and Vladivostok, and seasonal international flights.***

© Dan Weits / Shutterstock

146

Hit a spiritual gold jackpot at Yangon's Shwedagon Paya

MYANMAR (BURMA) // The 99m-high (326ft) *zedi* (stupa) of Myanmar's signature monument is plated with 22,000 gold bars and set with over 5500 diamonds and 2300 rubies, sapphires and other precious stones. At least 1000 years old, the *zedi* is the dazzling centre of a complex of shrines, statues, stupas and pagodas. A river of humanity flows around this gleaming hub, praying and leaving offerings to honour the Buddha. We defy you to visit and not feel a touch of the divine – particularly if you come at dawn or dusk, when it's bathed in a heavenly golden light.

SEE IT! ***The complex is in the centre of Yangon. Hire a guide to get the most out of the experience.***

147

Swim with the fishes and more at Ningaloo Marine Park

AUSTRALIA // The Great Barrier Reef gets all the attention, but discerning Australian aquanauts head west – just like the mighty migratory whale sharks that visit Ningaloo between March and July every year.

Unlike the GBR, which demands a long boat ride to see the best bits, it's possible to explore Ningaloo Reef straight from idyllic beaches in Cape Range National Park. All you need to see turtles, manta rays, dugongs and dolphins is a snorkel mask.

SEE IT! ***The main gateway towns are Exmouth and Coral Bay, 36km (22 miles) and 116km (72 miles) respectively from Learmonth Airport.***

148

Get an intimate perspective of the Eiffel Tower

FRANCE // Even before arriving in Paris, the Eiffel Tower is pre-etched in travellers' minds as the defining symbol of the city. And once you're here, it dominates the skyline at every turn. But it's not until you're within it that you truly appreciate this architectural innovation designed by Gustave Eiffel originally as a temporary showpiece for the 1889 World's Fair. After taking the lift to the 3rd floor, from where Paris spreads out at your feet, the best way to admire its wrought-iron lattice girders is descending the 720 stairs.

SEE IT! ***Plan to ascend the Eiffel Tower on a clear day; dusk offers day- and night-time views. Pre-purchase time slot tickets online.***

© Matt Kazmierski / Getty Images

Top: trace the trajectory of sliding stones at Death Valley's Racetrack Playa. Bottom: the Valley's endlessly photogenic Mesquite Flat Sand Dunes

149

Feel the heat in Death Valley National Park

USA // There's something invigorating and even life-affirming about being in an environment so punishing that even the rocks try to escape. Straddling California and Nevada, Death Valley claims the USA's hottest temperature (57°C/34°F) and also its lowest point (Badwater, 86m/282ft below sea level). While the name evokes a barren and lifeless place, the area actually supports plenty of native wildlife, from bighorn sheep to desert tortoises and even spectacular spring wildflower blooms. Eerie canyons, sand dunes that sing, boulders that mysteriously slide across the desert floor and extinct volcanic craters add to the spectacle.

SEE IT! ***Las Vegas is the closest major transport hub. The best time to go to Death Valley is November to April. Summer is crazy hot.***

© Checubus / Shutterstock

© Jonathan Gregson / Lonely Planet

© denizunlusu / Getty Images

150

Experience eerie volcanic energy on the Snæfellsnes Peninsula

ICELAND // A wild, windswept promontory inhabited principally by grazing horses, the Snæfellsnes Peninsula casts a strange and powerful spell on all who visit. Jules Verne set *Journey to the Centre of the Earth* here, and New Age mystics have hailed it as one of the planet's principle 'energy sources'. Even spiritual sceptics must admit it has a lunar feel: black lava fields dot the landscape, while lonely lighthouses count among the few signs of humanity, standing on storm-lashed cliffs, pods of orcas skimming by. Presiding mightily over all is the white mass of Snæfellsjökull herself – a glacier topping an active volcano, whose fiery innards featured in Verne's book. Take a snowmobile or guided hike to the top to be rewarded with Iceland's most epic views: beaches stretching into oblivion, glaciers gathering on the horizon and the surging tides of the Atlantic on three sides – all with barely a soul around.

SEE IT! ***A good base for exploring Snæfellsnes is Stykkishólmur, clustered around a natural harbour on the peninsula's northern shore.***

151

Indulge in the good life on Waiheke Island

NEW ZEALAND // Waiheke is a favoured weekend getaway for wealthy boho Aucklanders, and no wonder. The golden light slanting across the vineyards rivals Tuscany's, the beaches are near-tropical blue and the modernist clifftop homes would satisfy any tech millionaire aesthete. Spend a few days swanning around art galleries, sipping local Cabernets at wine bars, noshing on tapas at ivy-draped garden restaurants and strolling trails through nature reserves. In December, make your way to the island's bottom end, when pohutukawa trees bloom crimson over Man o' War Bay, one of Waiheke's primo swimming spots. For beachcombing, sunbathing, barbecuing and sandcastles, Onetangi Beach's 2km (1.2-mile) stretch of white sand is top-notch – there's even a discreet nudist section around the headland. Bikes are a great way to get around, but beware the hills!

SEE IT! ***Frequent ferries from Auckland take 40 to 80 minutes. Once there, rent a car or bike, or use the hop-on-hop-off bus.***

152

Ogle orchids and more at Singapore Botanic Gardens

SINGAPORE // The island state's sci-fi-style Gardens by the Bay may be superior eye-candy, but Singapore Botanic Gardens is no slouch in the beauty stakes and holds the ultimate trump card: World Heritage status. The accolade is richly deserved for this horticultural Eden which has been lovingly nurtured since 1859. Explore the grounds to find the world's largest showcase of tropical orchids, and don't miss the new Learning Forest where boardwalks take you across freshwater forest wetlands and an elevated walkway soars up to the rainforest canopy.

SEE IT! *Botanic Garden MRT goes to the Bukit Timah Gate, while several buses serve the historic Tanglin Gate, closest to Orchard Rd.*

© Igor G / 500px

153

Find a mix of moors, mines and culture in the Brecon Beacons

WALES // With its hundreds of square kilometres of beguiling peaks and moors and a Unesco World Heritage site (the ironworks town of Blaenavon). the rugged landscape of Wales' Brecon Beacons National Park may look wild, but it has been shaped by 8000 years of human settlement. Outdoor-lovers may just as likely chance upon Roman burial chambers and medieval castles as on waterfalls and caves. In its market towns, buzzing events such as the Hay Festival of Literature and Arts promise to continue the Brecon Beacons' rich cultural legacy.

SEE IT! *Take the train to Brecon, Abergavenny, Merthyr Tydfil or Llandovery. An Explore Wales Pass covers all rail and many bus routes.*

© Anthony Brown / 500px

© 52Hzwhale / Shutterstock

© Laura Storm / Alamy Stock Photo

154

Enter a wonderland of digital art at teamLab Borderless

JAPAN // The antithesis of the staid world of traditional art museums housing roped-off classics, teamLab Borderless is, as the name suggests, all about removing boundaries. Step inside a series of digital landscapes where the artworks react to your presence and transform as you move, touch, or simply stand still. The result of a collaboration between artists, animators, programmers and architects, each room is its own magical immersive experience: in one a virtual waterfall, in another an endless 'forest' of colourful lanterns, in the next a wall of flowers whose petals scatter when touched. Don't miss Crystal World, where you might encounter butterflies that have flown in from other artworks. The fun even continues to the tea room, where digital flowers bloom inside your teacup. Exhibits are ever-changing and there are no rules – come and play.

SEE IT! *teamLab Borderless is in Tokyo's Odaiba neighbourhood. Buy advance tickets online; arrive early on a weekday to avoid crowds.*

155

Dive into subaquatic eco-art at Mexico's Underwater Museum

MEXICO // Aquamarine with a heavy tint of 'green', this underwater sculpture park near Cancún is designed not just for aesthetic beauty, but with the intention of diverting human traffic away from endangered coral reefs nearby. Since opening in 2010, the Museo Subacuático de Arte (aka MUSA) has successfully tackled the issue of reef degeneration with creativity and flair, submerging around 500 sculptures at depths of up to 10m (33ft) in the waters off Cancún and Isla Mujeres. You can snorkel, dive or – for nervous swimmers or those averse to getting their feet wet – peer through the crystal waters in a glass-bottomed boat to view the fantastical creations, many of them created by British artist Jason deCaires Taylor. The highlight? A collection of 400 barnacled human statues lined up like terracotta warriors called 'Silent Evolution'.

SEE IT! *It's possible to organise snorkelling or diving trips to the museum using local operators in Cancún.*

156

Fall head over heels in love with Lisbon's Moorish Alfama

PORTUGAL // Whether you take a rattling ride on vintage tram 28 up the winding streets, almost grazing the front doors of azulejo-tiled houses, or wander on foot around its Moorish maze of back alleys, the Alfama enchants with its villagey vibe and hilltop *miradouros* (lookout points). The higher you go, the better the view across red rooftops and church spires tumbling down to the Tagus River.

The only neighbourhood still standing after the 1755 earthquake, the Alfama remains deliciously old-fashioned and low-key, with hidden plazas, laundry flapping and steep staircases leading to boho cafes, bars and restaurants where gentrification is slowly creeping in. By night, it takes on a more nostalgic air in dimly lit clubs that reverberate to the mournful sounds of fado, with guitar strumming and songs about the sea, fate and lost love.

SEE IT! ***You will get lost in the Alfama; accept that it's part of the fun. But take care at night and stick with others.***

Benedict Allen's Top Five Places

Author, environmentalist and film maker Benedict Allen is one of the world's leading modern-day explorers. He famously carries out his expeditions without a phone, GPS, or any back-up.

01

KHOVSGUL, NORTHERNMOST MONGOLIA – This isolated pocket of stark mountains and taiga forest is home to nomadic reindeer herders. It's a challenge in winter (when it can be -30°C), but come spring whole hillsides become pristine alpine meadows.

02

THE NAMIB DESERT, NAMIBIA – The only country named after a desert – and what a desert! Running up the dramatic 'Skeleton Coast', it transforms from wild seas of ochre sand in the south to lunar-like Damaraland.

03

RORAIMA, SOUTH AMERICA – Better known as The Lost World, this spectacular plateau rises from the remote forestlands where the borders of Brazil, Venezuela and Guiana meet – and withholds many of its secrets to this day.

04

CHUKOTKA, THE RUSSIAN FAR EAST – Not somewhere to go camping: it's -40°C during winter, which can last six or more months. But this is tundra at its best – bleak, treeless and breathtakingly stark. It's an utterly haunting landscape, one you'll never forget.

05

THE PERUVIAN AMAZON – Head out of Iquitos to the Peru-Brazil border and you'll find a labyrinth of moss-laden boughs filled with monkeys, parrots, sloths, anacondas – just like the Amazon of our imagination.

© Martijn Senders / EyeEm / Getty Images

© Dimitry Samsonov / 500px

157

Keep an eye out for wildlife in Khao Sok National Park

THAILAND // Ancient and vast, Khao Sok may be the rainiest location in the country, but it is also one of the rare slices of Thailand that are still habitable for large mammals. Its dramatic limestone terrain, with karsts shooting up as much as 960m (3150ft) and riven by cascading waterfalls, is a habitat for bears, boars, gaurs, tapirs, gibbons, deer, wild elephants and perhaps even a tiger among the juicy thickets.

A network of dirt trails snakes through the 739-sq-km (285-sq-mile) park: perfect stalking ground for fauna and flora – this is one of the most biodiverse areas of the planet, supporting some 200 different species. To encounter some of the aquatic ones, raft, canoe or kayak the Sok river, with its access to small water channels and mangrove swamps.

SEE IT! ***The nearest airports are Surat Thani and Phuket. Animal viewings are more likely in the wet season (June to October).***

158

Be floored by the depths of the Cañón del Colca

PERU // It's not just the vast scale of the Colca, the world's second-deepest canyon that buries its way through southern Peru, that makes it so fantastical. It's the shifts in its mood. There are more scenery changes along its 100km (62-mile) passage than in most European countries; from the barren steppe of Sibayo, through the ancient farm terraces of Yanque and Chivay, into the steep-sided canyon proper beyond Cabanaconde.

The most breathtaking canyon views are reserved for the condors gliding majestically overhead; land-bound humans must make do with dramatic viewpoints along the rim or a trek to the canyon bottom, where its depth – twice that of the Grand Canyon in the US – can truly be appreciated. You may bump into descendants of the Cabana and the Collagua people, who've called the passage home for centuries.

SEE IT! ***The town of Chivay, 160km (100 miles) from Arequipa, is the main starting point for trips to the Colca.***

159

Sail and snorkel the idyllic Bazaruto Archipelago

MOZAMBIQUE // Bazaruto Archipelago has been a getaway for island-hoppers in the know for decades, but it never loses its appeal. Lapping the white sands of the five main islands are clear turquoise and aqua waters filled with colourful fish, fat dugong and frolicking dolphins. On Bazaruto, the archipelago's largest island, flamingos wade in the shallows while sand dunes rise up in the background. This beauty comes at a price, and all of the archipelago's lodges are upmarket. Yet, if you decide to splurge, you'll have an unforgettable holiday.

☛ SEE IT! ***The islands are reached by boat from Vilankulo or via light aircraft from Vilankulo, Maputo or Johannesburg (South Africa).***

160

Shop in a 15th-century mall in the Grand Bazaar

TURKEY // Entering the Grand Bazaar, as centuries of browsers and Silk Road traders have done before you, greets your senses with the force of a commercial carnival. Stalls are piled with Turkish carpets, floral-patterned Ottoman tiles, *lokum* (Turkish delight), handcrafted jewellery, Uzbek silk and more. It can be overwhelming, but the shopkeepers are generally not as brash as their reputation suggests, and hagglers get a cup of *çay* tea. You could spend hours on these vaulted lanes, drifting between the caravanserais and cafes of this self-contained world.

☛ SEE IT! ***Open from Monday to Saturday, the Grand Bazaar is just off Divan Yolu Caddesi, the main road through Istanbul's old town.***

161

Walk, whoosh or wave-ride through stunning Snowdonia

WALES // Snowdonia National Park in North Wales brims with beauty and myth. Snowdon itself is a mountain for anyone – those who don't want to walk up can take a train, while hikers can follow numerous trails. But there are several other ranges in this extensive upland: try craggy Tryfan, with invigorating scrambling routes, or Cader Idris, home to legends of bottomless lakes and giants. Then there are the swathes of sandy beaches, the world's fastest zip line (Penrhyn Quarry) and a cutting-edge inland surfing lagoon (Adventure Parc Snowdonia).

☛ SEE IT! ***The Snowdon Sherpa bus service links the peak's trailheads, and trains run to Porthmadog, Harlech, Barmouth and Pwllheli.***

162

Swoon over the Uffizi's Renaissance superstars

ITALY // Dazzling, overwhelming, mind-expanding, exhilarating – how to describe a tour of this world-famous, historic, 101-room palace-museum? It is home to 1500-odd of the world's most dazzling masterpieces spanning five centuries, from the 13th to the 18th, bequeathed by the Medici family. As extraordinary as the Michelangelos, da Vincis and Botticellis is the luxurious setting: gilded and painted from top to toe with views down the Arno River from the Vasarian Corridor.

SEE IT! *Regular trains connect Florence to Pisa airport. Prebook Uffizi tickets online and collect them on arrival to avoid queueing.*

163

Chow down with the hospitable folk of the Connemara Peninsula

IRELAND // The starkly beautiful landscape of Connemara more than lives up to its name, the Irish word for 'Inlets of the Sea'. Tiny coves and jagged channels form a filigreed pattern along the peninsula's coastline and, inland, the scenery is no less dramatic, with desolate valleys, rusty bogs and green hills interspersed by black lakes. Village pubs host traditional music and serve seafood chowder made to closely guarded family recipes.

SEE IT! *It's easiest to get around by car, but hiking and cycling trails make Connemara a great place to explore on foot or by bike.*

164

© Yongyut Kumsri / Shutterstock

164

Cram in the culture at Berlin's blockbuster Museumsinsel

GERMANY // Berlin has not one or two but a whole Unesco-listed district of outstanding museums: Museumsinsel. Devote at least half a day or more to these vast collections, which sit astride the Spree Island and span a whopping 6000 years' worth of art, artefacts, sculpture and architecture. Highlights come thick and fast: the Pergamonmuseum for a fascinating romp through ancient Greece, Rome and Babylon; the Greek-temple-like Alte Nationalgalerie for 19th-century European art; and the Egypt-focused Neues Museum, where the bust of Queen Nefertiti reigns supreme.

SEE IT! *Save euros with a day ticket for all five museums. All collections are also covered by the Berlin three-day Museum Pass.*

© Matej Kastelic / 500px

With its slender columns and striated double arches, the Mezquita is one of the world's greatest examples of Islamic architecture

165

Discover the unity and harmony of Córdoba's Mezquita

SPAIN // Diversity reigns supreme in Córdoba's finest religious monument, a giant 1235-year-old mosque with a Christian cathedral plonked in the middle of it that sits on the cusp of the city's ancient Jewish quarter.

Back in the 10th century, in the days of the Umayyad Caliphate, Córdoba was a cultural colossus, the most advanced city in Europe and a centre for art, philosophy and learning. At its heart sat the indomitable Mezquita, once the most important Islamic monument west of Jerusalem but, today, an architectural anomaly and one of the only places in the world where you can worship Mass in a mosque.

In both scale and design the Mezquita is spellbinding. Enter via a 14th-century Moorish gateway into a fragrant patio replete with orange trees. The main body of the Mezquita is essentially one ginormous room held up by 856 slender columns supporting striped double arches and a roof decorated with multicoloured motifs. Random wanderings will take you past the priceless mihrab with its golden Byzantine mosaics and the beautifully incongruous cathedral with its jasper and marble altarpiece. Prepare to be awed.

SEE IT! *Córdoba lies between Seville and Granada in Andalucía. Weekday mornings are best for appreciating the Mezquita.*

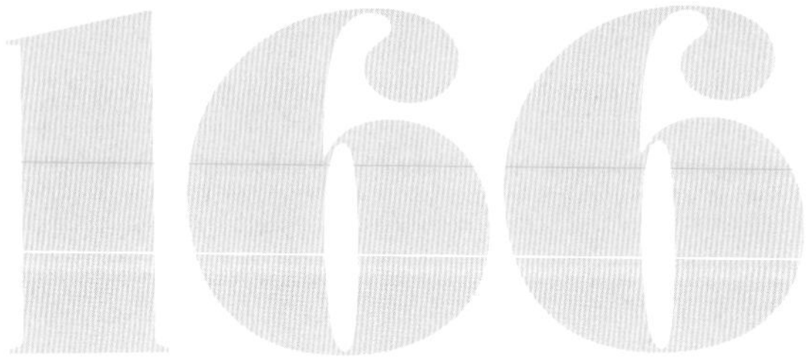

Top: desert winds sculpt the Sossusvlei dunes into rippling waves of ochre. Bottom: Sossusvlei silhouettes

Be beguiled by the hues and stark forms of Namibia's parched Sossusvlei

NAMIBIA // It is at Sossusvlei that the world's oldest desert is arguably at its most stunning. Sossusvlei is a large ephemeral pan, surrounded by dunes rising over 300m (985ft) high, sculpted by the wind into seemingly razor-sharp crests. At sunrise and sunset, the play of light is mesmerising, as the orange hues of the dunes are juxtaposed against the black shadows of the low-hanging sun. On the rare occasions when the Tsauchab River has enough water to fill the pan, the parched expanses turn into a striking blue-green lake that attracts abundant birdlife. Nearby in Deadvlei, the barren branches of ancient camelthorn trees make stark silhouettes against the white ground and blue sky.

SEE IT! ***Sossusvlei is a scenic 365km (227-mile) drive from Windhoek. Sleep inside the gates for sunrise or sunset access.***

© Samart Boonyang / Getty Images

167

Chill out at Wat Pho, Bangkok's Buddhist temple nirvana

THAILAND // If the enormous Reclining Buddha at gorgeous, colourful Wat Pho isn't enough to make you come over all Zen, how about an on-site massage? In addition to being one of Bangkok's biggest temple complexes, Wat Pho is also the national headquarters for the teaching and preservation of traditional Thai medicine, which includes Thai massage. Pavilions here facilitate that elusive yet wonderful convergence of sightseeing and relaxation (thank you Thailand, you're the best). The Reclining Buddha's pretty special, as well. Almost too big for its shelter, the genuinely impressive statue, which stretches out 46m (151ft) long and 15m (49ft) high, illustrates the passing of the Buddha into nirvana. It positively glows with gold leaf, while mother-of-pearl inlay ornaments the massive feet.

SEE IT! ***Take the Chao Phraya Express Boat to Tha Tien to reach Wat Pho, which is in Bangkok's Ratanakosin district.***

© Borisb17 / Getty Images

168

See how Ravenna's mosaics shine a light on the dark ages

ITALY // Ravenna's eight World Heritage sites give the lie to the myth that the Middle Ages were 'dark' and culturally dreary. Rome's decline was, in fact, Ravenna's opportunity as the city assumed the mantle of the capital in AD 402. The mosaics of the mausoleum of Galla Placidia twinkle so brightly they inspired American songwriter Cole Porter's song 'Night & Day'. Those in the Basilica di Sant'Apollinare are similarly dazzling, so much so that Pope Gregory the Great thought they should be blackened so as not to distract gawping congregants. Finer still is the Basilica di San Vitale, which glows with gold, emerald, amethyst and sapphire scenes from the Bible amid a gorgeous tangle of flowers, stars and birds. If you think looking around old churches is boring, come to Ravenna and prepare to rethink that.

SEE IT! ***Ravenna is 1½ hours from Bologna by train. Buy a cumulative ticket for all eight Heritage sites from the tourist office.***

© Matejh Photography / Getty Images

© Martin Hesko / Shutterstock

169

Get high on the otherworldly Danakil Depression

ETHIOPIA // Forget Virgin Galactic or the Mars One mission – you don't need to leave Earth to find another world. Simply board an Ethiopian Airlines flight and make your way to Mekele before starting your descent into the surreal depths of the Danakil Depression. With fumaroles spewing sulphurous gases, boiling pools of fluorescent brine and air temperatures often exceeding 50°C (122°F), this is somewhere to visit on a well-planned expedition. You'll plunge more than 100m (328ft) below sea level and find yourself at the heart of three diverging tectonic plates – this is one of the most geologically active places on the planet. Look out for lava flowing down Irta'ale Volcano, or a lake of molten rock in its crater. There are only two guarantees about an adventure here – you'll never forget it, and you'll sweat like there's no tomorrow.

SEE IT! *Expeditions to the Danakil Depression can be organised through tour operators in Addis Ababa and Mekele.*

170

Lose yourself in 145 rooms of fine art at the V&A

ENGLAND // The world's largest museum of decorative arts is full, as you'd expect, of grand and tasteful works from across the world and throughout history. Here you'll see giant sketches by Raphael, sumptuous carpets, Fabergé animals and an Elizabethan bed featured in *Twelfth Night*. But this palace of culture has an energy that belies its 150-odd years. Its 145 galleries house Nike Trainers, a selfie stick and '60s miniskirts as well as samurai armour and Rodin sculptures. Special exhibitions have taken in Christian Dior and David Bowie. The effect, as you stroll through corridors, past marvel after marvel, is appealingly democratic – everything here, from a miniature lock to a vivid animatronic tiger, has a weight, and deserves a second glance. There's too much for one visit, but it's free – so go, and come back.

SEE IT! *The museum is five minutes' walk from South Kensington Tube. There's a charge for special exhibitions.*

171

Disappear into the timeworn lanes of Old Québec City

CANADA // 'Old' in North America is relative. So Québec City's historic heart is all the more special. The city, founded in 1608, is the only one on the continent to have preserved its old defensive walls, gates and bastions. Inside these ramparts is a twisty, tumbledown, cobbled cluster of 17th- and 18th-century streets and houses more suited to France. Sitting at a sidewalk cafe with a vin rouge and plate of poutine (chips, cheese and gravy) is the best way to take it in.

SEE IT! ***Walking tours with maps, instructions and background info can be downloaded from www.quebec-cite.com.***

172

Hike the Lowlands to the Highlands on West Highland Way

SCOTLAND // You can take the low road or the high road on the West Highland Way, which begins on the outskirts of Glasgow and finishes beneath the looming mass of Ben Nevis. It passes Loch Lomond and Tyndrum, heads through vast Rannoch Moor, and runs along disused railways and drovers' tracks. The 155km (96-mile) route takes a week to walk, but it's possible to take on a single stretch, hop on a water bus en route, or detour to the ridges and peaks around it.

SEE IT! ***There is some accommodation en route, and it's possible to camp in places. Both Milngavie (where it begins) and Fort William can be easily reached by train.***

173

Tour the Pope's art collection at the Vatican Museums

ITALY // The Vatican museums are a city within a city. Founded in the 16th century by Pope Julius II, they hold 70,000 artworks (only 20,000 of which are displayed) telling the stories of Christianity through the ages. Artefacts from ancient Egypt, a golden gallery filled with maps and room upon room of paintings and frescoes, most famously by Raphael, leave the 6.5 million annual visitors dumbfounded. And, for the finale: Michelangelo's Sistine Chapel, which feels like the portal to heaven.

SEE IT! ***Vatican City is in Rome. The Museums are wildly popular: buy tickets online in advance and arrive as early as possible.***

Noo Saro Wiwa's Top Five Places

Noo Saro-Wiwa is an influential British/ Nigerian author and travel writer. Her book 'Looking for Transwonderland' details her return to Nigeria after her father Ken Saro-Wiwa, who campaigned against government corruption, was executed by the military dictatorship in 1995.

ETHNOLOGICAL MUSEUM OF BERLIN, GERMANY – This stunning collection includes some of the best African sculptures you'll find anywhere, including famous specimens from the Kingdom of Benin in Nigeria.

CONGO RIVER, BRAZZAVILLE, REPUBLIC OF CONGO – Sit on the banks of Africa's mightiest river and take in the sheer width, volume and power of one of the most unnavigable waterways in the world.

FOUTA DJALON HIGHLANDS, GUINEA – One of the most unsung but beautiful landscapes on earth. The view from the village of Telimélé offers a stunning panorama of tabular massifs rising from the forest floor.

MILLIONEN STERNEN HOTEL, AROSA, SWITZERLAND – A stellar experience, literally and figuratively. Guests spend the night in a sleeping bag on a viewing platform under the stars, and wake to a heavenly sunrise, views of endless peaks and the tinkle of cowbells.

FELA MUSEUM, LAGOS, NIGERIA – The home of Afro-beat music legend Fela Kuti is now this museum dedicated to his rebellious and outlandish life. His bedroom is as he left it (bed sheets and all), and on display are his saxophone, family photos and cuttings.

© Michael Wharley

174

Search for legendary submerged treasure at mighty Skógafoss

© Fred Concha / Getty Images

ICELAND // How can a waterfall stand out in this land of incredible waterfalls? Well, first, at 25m (82ft) wide and tumbling 62m (200ft) over a cliff at the western edge of the town of Skógar it's a sight to behold. Second, it's associated with a legendary tale. The first settler at Skógar, the slightly magical Þrasi Þórólfsson, stashed a chest of gold in the pool beneath the waterfall. For years the edge of the chest was visible, and many strong people tried to pry it free, but with no luck. Eventually, a group ended up tearing off the ring that attached to the chest, and the chest sunk away, never to be seen again. You can, however, see the ring in Skógar folk museum.

SEE IT! ***Skógafoss is two hours west of Reykjavík and is easy to reach by bus or car.***

175

Make your island dream a reality in beautiful Bora Bora

FRENCH POLYNESIA // A ring of sand-edged *motu* (islets) encircle a glinting turquoise lagoon around soaring rainforest-covered peaks: this dreamlike vision has made Bora Bora a byword for honeymoon luxury, augmented by its high-end bungalows, which stretch out like tentacles into the clear water. But there's more: divers and snorkellers can spot black-tipped sharks or stingrays among glittering shoals of fish. Hiking and parasailing are also available, while several affordable pensions and hotels mean it's just about possible to come here on a budget.

SEE IT! ***Reach Bora Bora by air or boat from Tahiti, 280km (174 miles) away. August to October usually sees dry and cool weather.***

177

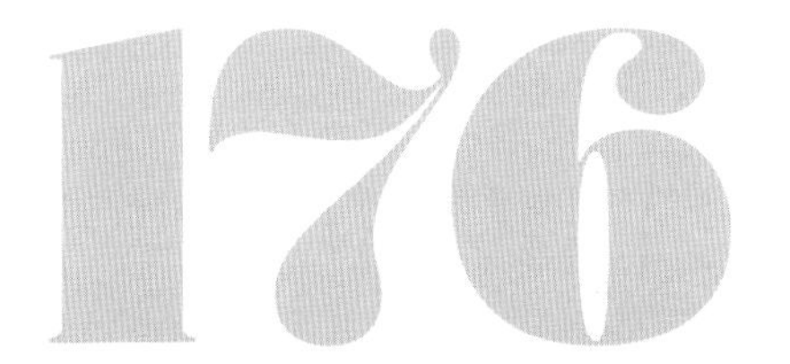

Idle a while in Isla de Ometepe, or get active on its peaks

NICARAGUA // There are myriad reasons to head to Nicaragua's natural objet d'art, Ometepe, the largest island in Central America's greatest lake. Volcanoes Concepción and Maderas dominate the island and are popular, with challenging hikes (go with a guide) taking you past howler monkeys and through cloud forests to the exposed summits. But Ometepe also offers rock carvings, beaches, kayaking and relaxation-inducing accommodation, not to mention a storied history of pirates and conquistadors and a reputation as a sacred place.

SEE IT! ***Boats run to Ometepe from San Jorge, a town 71km (44 miles) south of Granada and 98km (60 miles) south of Managua.***

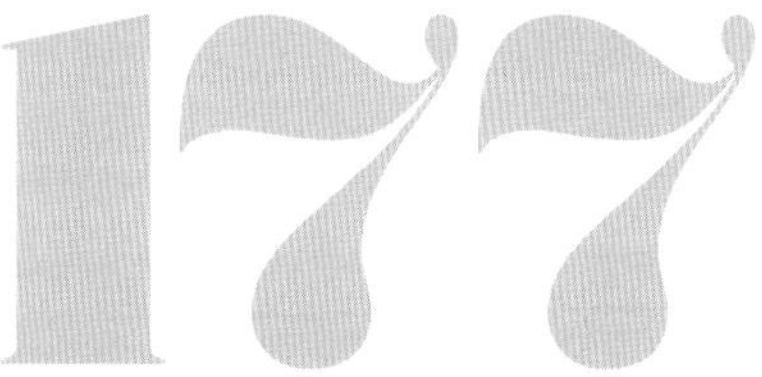

Explore the deep blue wonders of the Península Valdés

ARGENTINA // Once a hardscrabble peninsula of sheep ranches, today Península Valdés is known for some of South America's best marine wildlife watching. The main attraction is seeing endangered southern right whales get up close and acrobatic. Strictly regulated whale-watching tours visit these huge mammals from July to mid-December. But there are also killer whales (orcas), Magellanic penguins, sea lions, elephant seals, rheas, guanacos and seabirds. Shore walks feature lots of wildlife-watching opportunities, but dive or kayak for greater immersion.

SEE IT! ***Flights from Buenos Aires go to nearby Trelew or Puerto Madryn; there are also bus connections.***

© Matt Munro / Lonely Planet

© Anita Isalska / Lonely Planet

178

Step into the storied past of a Czech fairy tale at Prague Castle

CZECH REPUBLIC // Looming above the Vltava's west bank, this madcap montage of spires, towers and palaces dominates the centre of Czech capital Prague like a set from Cinderella and dates back to the 9th century AD. Within its walls lies a fascinating collection of historic churches, museums and galleries that are home to some of the Czech Republic's greatest artistic and cultural treasures. Pretty much every visitor who comes to Prague swings by, but what else would you expect from the country's answer to the Louvre? The world's largest ancient castle, its scale is exemplified in lavishness by the chapel of St (and one-time ruler) Wenceslas, and in sheer size by the Old Royal Palace, with its Rider's Staircase built sufficiently wide to allow knights to enter on horseback!

SEE IT! *Prague Castle is a steep uphill climb from Charles Bridge; tickets last two days, so don't rush to cram everything into day one.*

179

Slip, slide and glide through giant ice sculptures in Harbin

CHINA // In January, when the Harbin Ice & Snow Festival kicks off, average daytime temperatures don't rise past -13°C, perfect conditions for keeping the festival's magical array of ice sculptures intact. Yes, there are other winter sculpture fests in the world, but none of them manage to craft dozens of nearly life-size world landmarks out of ice. Many are structures you can walk into: stroll the arches of the Roman Colosseum, gaze up at a sparkling Eiffel Tower or roam through a slippery ice maze. Adding to the already over-the-top atmosphere is the fact that it gets dark early, and the sculptures are lit in a mesmeric array of colours: purple, blue, pink and magenta all creating a shimmering rainbow wonderland.

SEE IT! *The biggest sculptures are located at Harbin Ice & Snow World park on the west side of the island across the Songhua River.*

© Rudy Denoyette / Lonely Planet

© Michael Heffernan / Lonely Planet

180

Trace pilgrims' paths on the island abbey of Mont St-Michel

FRANCE // Outside the crowded summer season, tranquillity envelops Mont St-Michel, an ethereal abbey-island off the coast of Normandy. It's linked to the mainland by a 2014-built pedestrian and vehicle bridge designed to help preserve the island from sand and silt build-up.

Tidal ranges here famously vary by up to 15m (50ft); when they're low enough, you can take a spiritual bare-foot stroll across the rippled bay to the minuscule island accompanied by a guide (the waters rush in at speed that's said to be as fast as a horse in full gallop, so strolling on your own is not recommended). Centuries of Norman history seep through your soles as you trace the path of pilgrims who crossed these sands in the Middle Ages to reach the slender towers and sky-scraping turrets of the abbey, where Benedictine monks still hold regular services.

SEE IT! *Pontorson, 7km (4 miles) south of the La Caserne parking area, is the main rail hub, with shuttle buses to Mont St-Michel.*

181

Pay a visit to the Lord on high at Cristo Redentor

BRAZIL // What statue in the world boasts a view as sublime as Rio de Janeiro's Cristo Redentor? Standing atop Corcovado, arms outstretched in peace, this soapstone icon looks out over the city's spectacular landscape: a jumble of mountains and beaches, favelas and skyscrapers. It's a view almost as impressive for tiny mortals who ride up on the Corcovado rack railway and stand at its holy feet. Though plans for Cristo Redentor stretched back as far as the 1850s, it was only completed in 1931. Since then it has been struck by lightning; visited by Popes, presidents and kings; mischievously illuminated in the German national colours for the World Cup – but still claims its place in the starting line-up of Brazilian national treasures.

SEE IT! *If you plan to take the rack railway, arrive early or expect long queues – there are only limited train services daily.*

182

Holiday like a celebrity on Lake Como

ITALY // Lovely Lake Como was carved out of the foothills of the Alps by a mighty glacier which, once it receded, left behind an idyllic sapphire-blue lake framed by verdant green mountains. Its balmy microclimate, pretty Liberty villas, lavish gardens and wild hinterland have attracted the great and the good since Roman times. These days, George Clooney sips espresso in the village cafe. It's this balance of glamour and charm that makes Como so seductive. Pauper or prince, it will beguile you.

SEE IT! *Como is served by trains from Milan Malpensa airport and Milan Centrale. Once in Como use ferries to explore the lake.*

183

Stroll chameleonic chambers in Seville's Real Alcázar

SPAIN // One should be wary when making celestial comparisons, but Seville's Mudéjar-style royal palace, with its elegant salas and gardens, is certainly one of Spain's most heavenly buildings. A magnificent melding of Christian and Moorish design aesthetics, the Alcázar has experienced nearly a dozen rebuilds since its 10th-century inception, each leaving its own distinctive mark. The potpourri of styles served to inspire the producers of *Game of Thrones,* who used it to depict Dorne.

SEE IT! *High-speed trains connect Seville to Madrid (two hours). Avoid June to August when temperatures can be fierce.*

184

Join an extreme scramble at Shibuya Crossing

JAPAN // On the admittedly very short list of notable pedestrian crossings of the world, Tokyo's Shibuya Crossing is one many may not be able to name but have probably seen pictures of. This is thought to be the world's busiest intersection, where five roads meet and hundreds of people (upwards of 3000 at peak times) cross when the lights change. All around is the Tokyo of the movies – neon canyons, giant video screens – an electric setting for a massive human spectacle. It's best at night, particularly on Fridays or Saturdays when crowds spilling out of the station are at their thickest.

SEE IT! *Mag's Park, on the rooftop of the Shibuya 109-2 department store, is the top spot for overhead views of the action.*

© r.nagy / Shutterstock

© Jess Kraft / Shutterstock

185

Join thousands in pilgrimage at St Peter's Basilica

ITALY // This mighty basilica is erected on the spot where Rome's first Pope, the Apostle Peter, was crucified by Nero. The vast church that surrounds his tomb was built by a string of star architects, including Bramante, Michelangelo, Maderno and Bernini in the 16th century, and it remains the largest, wealthiest church in Europe, perched atop Vatican Hill like a beacon for the faithful. The cavernous interior is equally opulent, layered in marble and filled with artworks, including Michelangelo's moving Pietà. Why not attend a papal mass with 14,999 other worshippers?

SEE IT! *Free, two-hour English-language tours are run by the Pontifical North American College. Dress appropriately (no shorts, miniskirts or bare shoulders).*

186

See ruins reclaimed by nature in jungle-shrouded Palenque

MEXICO // In a sublime jungle setting in tropical Chiapas, Palenque is, arguably, Mexico's most atmospheric ruins, courtesy of the verdant surroundings, swirling morning mists and richly detailed Mayan architecture. More ancient and less crowded than Chichén Itzá, it's possible to commune with both history and nature here. Howler monkeys and toucans populate the nearby jungle while the ruins are so embedded in their setting that British novelist Graham Greene once described them as: 'so age-old they have a lichenous shape and look more vegetable than mineral'.

SEE IT! *Palenque town is 8km (5 miles) from the ruins but, for more atmosphere, head to charming El Panchán, a forest retreat between the town and ruins.*

© Milonk / Shutterstock

Presided over by the magnificent Shah Mosque, Unesco-listed Naqsh-e Jahan is one of the largest squares in the world

187

Spot the deliberate mistakes in Esfahan's Naqsh-e Jahan Square

IRAN // At the heart of Esfahan, the city that locals still declare is 'half the world', lies the mighty Naqsh-e Jahan, or Imam Square. Built in 1602 as the centrepiece of Shah Abbas' new capital, the square was designed to contain the finest architectural jewels of the Safavid Empire. On its south side stands the Shah Mosque, a melange of domes, vaulted arches and distinctive blue-tiled mosaics that remains unblemished after more than 400 years. Its entrance portal looms more than 30m (98ft) high and swirls with calligraphy and floral ornamentation. Look closely and you'll find deliberate mistakes, meant to show that only Allah can produce perfect work. Through the entrance, a short corridor was required to connect the alignment of the square with the direction of the mosque, which is orientated towards Mecca.

The smaller, but no less magnificent, Sheikh Lotfollah Mosque occupies the east side. This rhapsody in blue-patterned tiles has some of the best-surviving mosaics from the Safavid era, which are topped by fine *muqarnas*, highly decorated, stalactite-like carvings that seem to drip from the ceiling. The mosque's inner dome is a masterpiece of design; sunlight radiates through and shimmers on the ceiling's gold tiles to quite an effect.

On the western side lies the palace of Ali Qapu, the residence of the visionary shah. At the crown of the six-storey mansion is a covered terrace with slender wooden columns, a delicately painted roof, and panoramic views of the entire square and the network of bazaar arcades that connects these monumental structures. The oldest parts of the bazaar predate Shah Abbas' ambitious square expansion and date back more than a thousand years.

SEE IT! ***In early evening, Naqsh-e Jahan Square comes alive with picnicking families and promenading couples. Earlier in the day, keep cool under the marketplace arches surrounding the square.***

© Peter Etchells / Shutterstock

188

Pass through Platform 9¾ to the Wizarding World of Harry Potter

USA // Universal Orlando Resort puts you smack into the story, bringing Harry Potter's realm to life in exquisite detail. Poke along the cobbled streets of Hogsmeade, sip frothy Butterbeer and munch on Cauldron Cakes, all in the shadow of Hogwarts Castle. The authenticity tickles the fancy at every turn, from the screeches of mandrakes in the shop windows to the groans of Moaning Myrtle in the bathroom. And the experiences keep on coming: Feel the cold chill of Dementors, soar over the castle in a Quidditch match, visit Gringotts Bank with its goblin tellers. The Wizarding World is split between two parts of Universal's complex: Hogsmeade, at Islands of Adventure, and Diagon Alley, at Universal Studios. The Hogwarts Express train (naturally) links them.

SEE IT! *The Wizarding World is in Orlando, Florida. Early September, early November and May see the lightest visitor numbers.*

© MB Photography / Getty Images

189

Plunge into the life aquatic in the Bay of Islands

NEW ZEALAND // Māori were already ensconced in the Bay of Islands well before Europeans sailed in and claimed the area as the 'birthplace of the nation'. It was the site of New Zealand's first capital (1840–41), and the place where the Treaty of Waitangi – the country's founding document – was signed.

Its history is certainly interesting, but natural beauty is what makes the Bay so appealing. Around 150 undeveloped islands are sprinkled throughout harbours and hidden coves awash with shimmering turquoise waters, which are rich in marine life. It's brilliant for an oceanic outing, with the signature trip being a Hole in the Rock cruise. Among a raft of other options are fishing charters, kayaking, snorkelling and scuba diving.

SEE IT! *The Bay of Islands is around three hours' drive from Auckland. Historic Russell is the preferred base for its pretty setting.*

190

Hear the echoes of Mozart at Vienna's Schloss Schönbrunn

AUSTRIA // If ever a building could sum up an empire, Schloss Schönbrunn would surely be it. All the pomp and glory of the 600-year Hapsburg reign is evoked in this 1441-room summer palace. First up should be a spin around the lavishly frescoed, gilded and chandelier-lit interior (some 40 rooms are open to the public). Among the most spectacular are the 19th-century private apartments of Franz Josef I and his beloved wife Elisabeth, and the grand Mirror Room, where a six-year-old Mozart gave his first performance to a rapturous Empress Maria Theresa.

After this baroque feast, stroll the French formal gardens, a royal playground of mythological sculptures, follies, mock Roman ruins and flower beds. The hilltop gloriette affords cracking views over the palace to Vienna beyond.

SEE IT! ***Metro line U4 runs to Schönbrunn. The palace gets very busy during peak months. Tickets are stamped with a departure time so buy yours straight away and explore the gardens.***

© Manfred Kostner / awl-images.com

Find paradise in the turquoise and white palette of Mo'orea

FRENCH POLYNESIA // Charles Darwin described it as 'a picture in a frame'. And Mo'orea still looks like a perfect work of art: the deep blue of the Pacific wrapping around a turquoise lagoon, which itself encircles emerald cliffs and islets. Yet this laid-back volcanic island is far from one dimensional. It's big enough to absorb the tourists who flock here for a slice of Pacific perfection, and its lush peaks offer some serious hiking opportunities. Lavish villas and modern hotels are available alongside more basic bungalows, and there's a good range of independent restaurants, making Mo'orea one of the best bases in the region. The lagoon, with its gleaming white sandbanks, is simply blissful. Take a boat trip for the chance to spot stingrays, humpback whales and spinner dolphins.

SEE IT! ***Less than 20km (12 miles) of Pacific separates Tahiti and Mo'orea, which can be reached by air or boat. May to October is the driest time; whale-watching season runs July to October.***

© Marconi Couto de Jesus / Shutterstock

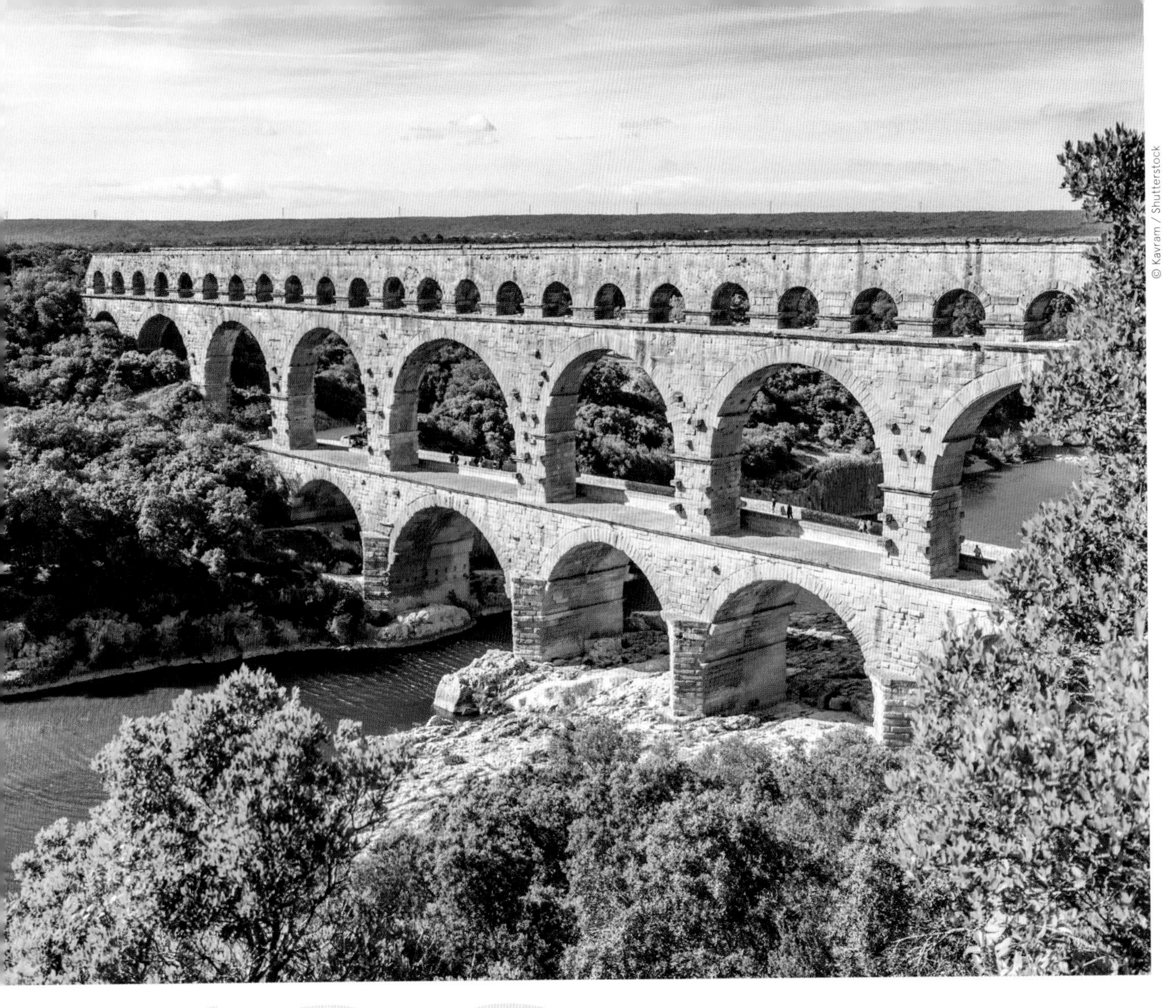

© Kavram / Shutterstock

Built by the Romans around 19 BC, the Pont du Gard is an astonishing feat of engineering

192

Marvel over (or under) a Roman engineering marvel at Pont du Gard

FRANCE // The Romans knew a thing or two about grandiose engineering. Their exceptionally preserved three-tiered aqueduct Pont du Gard in Languedoc formed part of a system of channels 50km (31 miles) long built around 19 BC to transport water from Uzès to Nîmes.

Archaeologists have since unearthed the extraordinary technicality and precision of the Romans' endeavour: the soaring aqueduct's hand-carved, locally quarried rock incorporated numbered stones, scaffolding support and the use of hoists. It descends by 2.5cm (1in) across its length, providing the necessary gradient to keep the water flowing across the river. Using nearly one thousand men, the colossal structure was, astonishingly, completed in just five years.

Floodlit in the early evening, the aqueduct straddles two bushy banks with sandy river beaches and a walking trail nearby. You can walk along the tiers, but to truly grasp its sheer scale – a whopping 275m (900ft) long and 48.8m (160ft) high – paddle out in a canoe or kayak to gaze up in awe from the water. April, May and June are best, as winter floods can make the river impassable.

SEE IT! *Buses between Nîmes and Uzès stop near Pont du Gard. Canoe hire companies are a two-hour paddle away in Collias.*

© Matt Munro / Lonely Planet

© Saiko3p / Shutterstock

193

Give yourself over to the alluring chaos of Hanoi Old Quarter

VIETNAM // Hanoi's Old Quarter is the complete Indochine package – a little bit French, a little bit communist and a whole lot Vietnamese. Here, French colonial mansions rub chopsticks with frenetic Southeast Asian street markets and cool cafes. Shop-houses are stacked up crazily like boxes in a warehouse and storefronts spill out into the street in flowing reams of rainbow-coloured silk, leaving just enough room for a constant stream of weaving motorcycles and wandering street-food vendors. Dotted amid the chaos are historic treasures, cheap hotels and some spectacular places to munch on everything from pho (noodle soup) to banh mi (baguettes). Sure it's commercial – as Hanoi's main retail district, that's rather the point – but it's hard not to fall in love with its unbridled exuberance and brash joie de vivre.

☛ SEE IT! *Hanoi is Vietnam's northern capital; a stroll through the Old Quarter can last from an hour to a day.*

194

Go with the literary flow along the Romantic Rhine

GERMANY // Flowing through some of Germany's most lyrical landscapes, the Rhine croons romance like no other river, especially as it twists and turns between Rüdesheim and Koblenz. This is the stretch that inspired those old Romantics to whip out their paints and pens – Turner, Byron, Goethe, Wagner and Shelley included. They thrilled to the sight of robber-baron castles, near-vertical terraced vineyards, dramatic cliffs rearing above churning whirlpools, and half-timbered villages straight from a Grimm fairy tale. You can go down the cruise or boat trip route, but the Romantic Rhine truly casts its spell when you sidestep the crowds. Hiking trails thread through quiet woods and vineyards to giddy lookout points, ruined fortresses and taverns serving local wines. Or stick closer to the river on the Rhine Cycle Route.

☛ SEE IT! *The Romantic Rhine is 100km (62 miles) from Frankfurt. One-way hikes can be combined with ferries, trains and buses.*

195

Embrace fire and ice in Vatnajökull National Park

© Matt Munro / Lonely Planet

ICELAND // If your inner explorer has ever dreamt of conquering the Arctic, you'll love Vatnajökull – a megapark of glacial ice and volcanic landscapes established to protect the world's largest ice cap outside the poles. Formed by the merging of two parks in 2008, the massive reserve covers 14% of Iceland. The northern portion of the park, Jökulsárgljúfur, protects a unique subglacial eruptive ridge and 30km (19-mile) gorge, as well as the grand waterfall Dettifoss. The southern part is anchored by Skaftafell, a wildly popular collection of peaks and glaciers. Though visitors flock here, so vast is the wilderness that there's always a trail to call your own. Head for the famous basalt waterfall, Svartifoss, in the shadow of brilliant blue-white Vatnajökull, with its lurching tongues of ice, and strike out from there.

☛ SEE IT! ***Jökulsárgljúfur is 500km (310 miles) from Reykjavík.***

Geoffrey Kent's Top Five Places

Worldwide travel pioneer Geoffrey Kent is the founder and CEO of Abercrombie & Kent, an international luxury travel company operating tours by air, land and sea.

ANTARCTICA – One of earth's last remaining natural frontiers, its hostility, isolation and sweeping beauty command respect, and its snowy expanse and towering sheets of ice ringed with ocean provide a wondrous natural habitat for a wide range of wildlife.

KENYA'S MASAI MARA – The scene of one of the planet's most dazzling displays of wildlife in the annual Great Migration, when millions of wildebeest and zebra search for greener pastures, while predators wait their chance.

INDIA – India offers a lifetime of experiences in one captivating country, from the snow-capped mountains of the high Himalaya and the deserts of Rajasthan to the jungles of Madhya Pradesh and the beaches of Kerala.

EGYPT – With unparalleled cultural and archaeological heritage, it's been attracting visitors for millennia. Whether it's Giza's monumental pyramids, Luxor's colossal Valley of the Kings, or the incomparable river Nile, this country is generous with its riches.

BRAZIL – Ipanema and Copacabana beaches may be on everyone's fantasy list, but there are many others. Rio De Janeiro is a must, but don't miss the colourful 16th-century Portuguese colonial towns in the north – and Florianopolis is another hidden treasure.

© Justin Weiler / Abercrombie & Kent

© Kevin C Moore / Getty Images

© pilesasmiles / Getty Images

196

Trace Unesco's logo in the Temple of Concordia

ITALY // On seeing the Temple of Concordia, you may be struck with déjà vu. It's so familiar, surely you've been before? But how could you forget a perfectly preserved, 2000-year-old Greek temple set on a ridge looking out at a gently ruffled sea? It's simply that you've seen it a dozen times on Unesco's logo. In Agrigento's Valley of the Temples there is a whole string of such beauties lined up on a ridge. At dusk they glow rose gold in the warm southern sunset, guiding sailors home.

SEE IT! *Trains run to Agrigento from Palermo. City bus 1 departs from the station to the entrance, where you can pick up audio guides.*

197

Take a cableway to heaven on the Sea to Sky Gondola

CANADA // On a balmy evening with a steep but satisfying climb behind you and a bowl of poutine in front of you, there are few better places to enjoy the pine-scented natural essence of British Columbia than from the top station of Squamish's Sea to Sky Gondola. The glorious glass-and-wood Summit Lodge looks out over the craggy Stawamus Chief and a 100m-long (330ft) pedestrian suspension bridge is slung over a gaping chasm behind. Rest your knees and take the gondola down.

SEE IT! *Squamish is an hour's drive from Vancouver. Buy Gondola tickets at the base station off Hwy 99, 4km (2.5 miles) south of town.*

Opposite, from top: Nara's giant Buddha is one of the world's largest bronze statues; take to the air for the best view of the mysterious Nazca lines

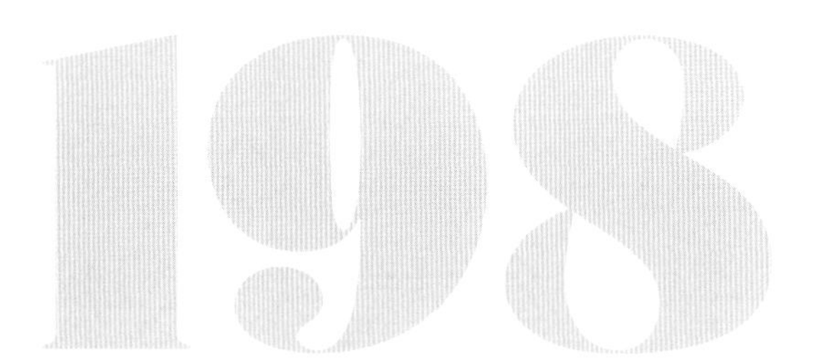

Seek enlightenment at the Daibutsu (Great Buddha) of Nara

JAPAN // The world is not short of statues of Buddha, but, if you're going to see just one, the Daibutsu of Nara has to be a top contender. Cocooned inside the Daibutsu-den (Great Buddha Hall) at temple Tōdai-ji, this image of the cosmic Buddha is one of the world's largest bronze figures, towering 15m (50ft) high and weighing some 500 tonnes. Its gracefully raised hand alone is the height of a person. Imagine the reverence this must have provoked when it was first unveiled in 752. Though parts of the statue have been recast over the centuries, it has lost none of its effect. The hall that houses the statue is necessarily immense, and is one of the world's largest wooden buildings (amazingly, only two-thirds of its original size). Towards the back you'll see a wooden column with a hole through its base. Popular belief maintains those who can squeeze through the hole, which is the same size as one of the Great Buddha's nostrils, are ensured of enlightenment – reason enough to pay a visit. Enlightenment secured, explore the leafy expanse of park Nara-kōen, where the deer rule.

SEE IT! *Nara is in the region of Kansai, about an hour by train from Kyoto and Osaka.*

Soar over the mesmeric desert mystery of the Nazca Lines

PERU // One of the greatest, most widely discussed archaeological mysteries, the ancient Nazca Lines comprise more than 800 straight lines, 300 geometric figures, and 70 animal and plant drawings spread over 500 sq km (193 sq miles) of arid, rock-strewn plain in Peru's Pampa Colorada. Favourite figures include a monkey with an elaborately curled tail, a hummingbird, a bizarre owl-headed person often referred to as an astronaut because of its goldfish-bowl-shaped head, a 180m-long (590ft) lizard and a condor with a 130m (427ft) wingspan. Some say these mystical beasts look better in photos, but do witness this stunning spectacle with your own eyes. Gawking at these gargantuan geoglyphs is possible either from a viewing tower beside the Carretera Panamericana Sur, about 20km (12 miles) from Nazca, or by flying over them for optimum views. The sight is enhanced by the puzzling matter of how the Nazca Lines could have been created so long ago, given they can barely be discerned from ground level and only properly seen from the air. This has led to countless debates about their age and original purpose.

SEE IT! *Dozens of companies offer flights over the lines from Maria Reiche Neuman Airport. The ride can be bumpy.*

198

199

200
299

Seek movie-set views atop the Empire State Building

USA // This art deco sky-tickler in downtown New York City is the world's most famous building – if we judge by Hollywood close-ups. It has starred in around 100 films, from *King Kong* to *Independence Day*. Heading to the top is a quintessential American experience. First, you need to make a choice: ascend to the open-air deck on the 86th floor, where wind tickles skin, and sounds of the sleepless city float ever so faintly in the air, or keep climbing to the 102nd floor, where a smaller, enclosed deck awaits. Both provide sparkling views over the Big Apple. Photographers and romantics prefer the tactile lower experience; superlative seekers opt for the higher ground (though you have to splash more cash to upsize).

SEE IT! ***Viewing platforms open 8am–2am. Go very early or late, or buy tickets online, to avoid horrendous queues.***

© Gary Latham / Lonely Planet

Reach new heights while climbing the Tsingy de Bemaraha

MADAGASCAR // In a country renowned for its wildlife – lemurs, chameleons, aye-ayes, fossas – it takes something special to steal the show. Presenting.... Tsingy de Bemaraha. Hewn by wind, rain and the hands of time, these fantastical-looking karstic limestone formations in western Madagascar are as astounding to behold as they are tricky to navigate. The local word for the razor-sharp pinnacles is *tsingy*, which translates as 'the place where one can't walk'. Yet you'll find yourself squeezing through cavities and reaching spiky summits hundreds of metres up. How? With the aid of trained guides and a via ferrata system of fixed cables, stemples, ladders, suspension bridges and walkways. If you're lucky, you may just spot a lemur or two in the forests below.

SEE IT! ***Camions-brousses (4WD army-style trucks) run from Belo-sur-Tsiribihina every few days in dry season (May to October).***

© Justin Foulkes / Lonely Planet

© Matt Munro / Lonely Planet

202

Take a front-row seat for glacial theatrics at Glaciar Perito Moreno

ARGENTINA // Among Earth's most dynamic and accessible glaciers, Perito Moreno is the centrepiece of the southern sector of Parque Nacional Los Glaciares. Branching off the Southern Ice Field, this creaking blue-tinged behemoth is 30km (19 miles) long, 5km (3 miles) wide and 60m (200ft) high. What makes it exceptional is its constant advance: up to 2m (7ft) a day. Its slow creep builds suspense and pressure until building-sized icebergs calve off the face and crash into Lago Argentino with ear-shattering cracks. Watching is a sedentary park experience that manages to be thrilling. A network of steel catwalks and platforms bring you face to face with the glacier, while lake cruises on Argentina's largest lake offer an aquatic perspective.

SEE IT! *Perito Moreno lies 80km (50 miles) west of El Calafate, with air and land transport connections.*

© Peter Johnson / Corbis / VCG / Getty Images

203

Experience the desolation of Namibia's Skeleton Coast

NAMIBIA // Countless ships have met their end on this foggy stretch of current-crossed Atlantic shoreline. Of those seafaring visitors who did reach land, many more perished amid the barren expanse to the east. Travellers today are captivated by the scale and majesty of the seascape, with its crashing waves, sculpted dunes, rusting shipwrecks and bleached whale skeletons. Sharks patrol the depths, seals dominate the shallows, cormorants perch on the wrecks. Then there are the elephants and lions roaming the dry river valleys. The coastline becomes remoter and more beautiful the further north you go. Seeing the southern section is straightforward enough under your own steam, but to explore the north you will need to arrange a specialist safari.

SEE IT! *Approach via the coastal road from Swakopmund, or from the north via the rugged Kaokoveld and Damaraland.*

204

Tread the monumental streets of Teotihuacán

MEXICO // While Egypt has the pyramids of Giza, Mexico harbours the temples of Teotihuacán. Two different cultures on two continents, both ambitiously innovative and way ahead of their time. Even as Mexican ruins go, Teotihuacán is humongous. At its high-water mark, up to 250,000 people lived in this energetic city. Today, four million a year stroll in awe through its haunting remains, along the 2km (1.3-mile) Avenue to the Dead from the ornately carved Temple of Quetzalcóatl to the peerless Pyramid of the Sun, third-largest pyramid in the world.

SEE IT! *The pyramids lie 50km (30 miles) northeast of Mexico City. Avoid 10am to 2pm, when crowds are thickest.*

205

Walk between marble cliffs at Taroko Gorge

TAROKO NATIONAL PARK // Named for the local Truku indigenous tribe, this canyon is a highlight of mountainous Hualien. The 18km-long (11ft) ravine was formed over millions of years as the steel-blue Liwu River carved its way through a slab of marble and gneiss to empty into the Pacific. There's a mystique to Taroko's misty cliffs; when a patch of fog lifts, you're liable to spot a hidden-away pagoda or rock-carved shrine. Plus, you can have fun donning a safety helmet and trekking through low- hanging rocks, next to waterfalls above the rushing Liwu.

SEE IT! *A tourist bus goes from county capital Hualien. A few hikes in the national park require permits, best arranged via a tour agency.*

206

Learn to climb Railay's dramatic limestone walls

THAILAND // If you've ever fancied yourself clambering up rock faces with the agility of Spiderman, then Railay (also spelt Rai Leh) is a superb place to learn the ropes. Blessed with towering limestone cliffs fringed with lush jungle and perfect beaches, this is one of the world's premier rock-climbing locations. You'll find myriad routes, ranging from relatively easy to super-challenging. In recent years Railay has become busier with day trippers and noisy boat traffic, but it remains far less developed than Ko Phi-Phi.

SEE IT! *Long-tail boats and/or ferries run to Railay from Krabi's Tha Khong Kha, Ao Nang, Ao Nam Mao, Ko Phi-Phi and Phuket.*

© Shaun Jeffers / Shutterstock

© Naughty Nut / Shutterstock

207

Be dazzled by a galaxy of glowworms at Waitomo Caves

NEW ZEALAND // Proving you don't have to rocket into space to discover alien landscapes, Waitomo Caves are all the more surprising for their location beneath workaday New Zealand farmland.

There are various portals into this underworld. At Ruakuri Cave, a futuristic stairway corkscrews down into labyrinthine passageways and vast caverns. Nearby is Glowworm Cave, with its acoustically sweet, cathedralesque chamber, and the chance to cruise beneath a 'sky' twinkling with a galaxy of glowworms – the luminescent larvae of the fungus gnat.

It's all rather thrilling, but for the ultimate deep-space exploration, embark on a legendary Black Water Rafting trip, where you get rubbered-up with wetsuit and inner tube, ready to clamber through cavities and float along subterranean waterways.

SEE IT! ***Waitomo is less than three hours' drive south of Auckland, in reach of other North Island highlights, such as Rotorua and Taupo.***

208

Cross Prague's time-spanning Charles Bridge

CZECH REPUBLIC // If Charles Bridge was still the Czech capital's main artery, Prague would have gone purple and died years ago. Fortunately, there are alternative ways across the Vltava River into the celebrated City of a Hundred Spires now, and the historic thoroughfare is pedestrianised: good news for the sightseers wishing to gawp at this elaborate structure and the river views it offers. At 520m (1700ft) long, with 16 arches and 30 baroque statues, the bridge had its foundation stone laid by Charles IV in 1357, on 9 July, at 5.31am. Exactly. The Holy Roman Emperor, a numerologist, allegedly selected this precise moment because it forms a numerical bridge, or scale: 1-3-5-7-9-7-5-3-1. Watch out for the Bradáč (Bearded Man), a stone head at the Staré Město end. Traditionally, when the river rose above the Bradáč, locals legged it into the hills.

SEE IT! ***View the bridge from a different angle during a boat trip along the Vltava.***

© Joshua Small / Shutterstock

© Photosounds / Shutterstock

Plunge into outdoor adventure at Lake Wanaka

NEW ZEALAND // Wanaka is New Zealand's alternative adventure hub – somewhere you can dive into all the outdoor activities the South Island is famous for, without surfacing to the high-octane beat of backpackers on an adrenaline rush. It's a smaller, cuter Queenstown, with the volume dial tweaked. At the lake's southern end, this is the gateway to mighty Mt Aspiring (NZ's highest peak outside the Aoraki/Mt Cook region) and the snow-sport centres of Treble Cone, Cardrona, Harris Mountains and Pisa Range. Off-piste pursuits include tramping, mountain biking, climbing, canyoning and kayaking. And, yes, snap a picture at the Instagram-famous Wanaka Tree growing in the lake – but don't climb! Careless visitors have imperilled the delicate willow.

SEE IT! *Queenstown Airport is an hour's drive away, with flights from all over New Zealand and Australia.*

Look back into ancient history at Masada National Park

ISRAEL & THE PALESTINIAN TERRITORIES // Surveying the hazy horizon from this desert plateau can blur your sense of time and place. King Herod bulked up the fortress here between 37 and 31 BC, building two palaces; reconstructed mosaic floors and the remains of synagogue walls testify to its early days. According to accounts by Roman-Jewish historian Josephus, dagger-wielding Jewish freedom fighters, known as the Sicarii, seized the fort after Jerusalem's conquest by Romans in AD 70. When they realised the enemy was about to breach the gates, the Sicarii died by mass suicide rather than face enslavement by the Romans. Their legendary last stand makes this arid expanse of fragmented history and weather-beaten ruins even more compelling.

SEE IT! *A cable car hoists visitors to the park. Or start early and walk Snake Path (Dead Sea side) or Roman Ramp Trail (west side).*

© Philip Lee Harvey / Lonely Planet

Pilgrims near the cross-shaped church of Bet Giyorgis, hewn from the rock in the late 12th or early 13th century

211

Descend into Lalibela's subterranean world frozen in stone

ETHIOPIA // Right across this list, great monuments feature that have been built by stacking stone on top of stone – making structures higher, bigger and more elaborate than ever before. Perhaps the greatest exception is Lalibela – Ethiopia's foremost historical site, and a city that follows an entirely different rulebook. The 11 monolithic churches that make up this holy city haven't been built in the conventional sense, but rather quarried out of natural bedrock – their forms chiselled into existence in the 12th and 13th centuries. It's not only the extraordinary structures that make this such a compelling place to visit. The city is a headquarters of the Ethiopian Orthodox Church – one of the world's oldest strands of Christianity, tracing its history back to the days of Solomon and the Queen of Sheba – and pilgrims walk from across the country to pray here. There are few experiences more humbling than joining them as they tread barefoot through the subterranean tunnels on the final stretch of their journey, or seeing white-robed congregations pray as centuries-old frescoes of saints peer down from the shadows above.

SEE IT! ***Be sure to watch sunrise from the grassy hill overlooking the famous church of Bet Giyorgis.***

Below: looking down from Meall Mor hill, whose flat-topped summit gives magnificent views of the giants around Glen Coe

Find stark beauty and bloody betrayal in Glen Coe

SCOTLAND // Think spectacular mountain scenery and you might imagine alpine villages or fairy-tale forests. Well, Glen Coe is nothing like that: it's just a desolate expanse of rock, heather and thistle. Cosy it is not, but this is a land of rugged poetry. Its boggy lower reaches are spotted by trees, moving up past waterfalls to snow-dusted Munros that rise sheer along the valley.

Yet this apparently barren wilderness has stories aplenty. It's home to legends of Celtic heroes and cattle rustlers, and was the site of the massacre of Glencoe (named after the glen's main settlement), where 38 members of Clan McDonald were massacred by British forces who had been billeted with them. Today, winter hikes, scrambling routes, wild camping spots and the Glen Coe ski area pull in the visitors. However you trace this Highland glen's legendary topography, its weatherworn beauty is not easily forgotten.

SEE IT! *Buses between Glasgow and Uig (Skye) stop at Glencoe village.*

© Tami Freed / Shutterstock

© Lucidio Studio Inc / Getty Images

213

Defy the depths in the Blue Hole

BELIZE // As every mountaineer must ascend Everest, so too are scuba divers drawn to the depths of Belize's Blue Hole Natural Monument. Jacques Cousteau charted the hole's depth at 124m (407ft) in 1971, declaring it one of the world's top scuba sites. From above it looks like a perfect, 305m (1000ft) blue circle but when divers drop in, they descend past ancient, storeys-high stalactites hanging from the roof of the underwater cave. The lucky ones catch a glimpse of a Caribbean reef shark or two, and maybe even a hammerhead. From the bottom of the hole, light from above is but a pinprick, towards which divers resurface slowly and with great care. Snorkellers can glide around the reef-covered rim, encountering small fish and the occasional turtle.

SEE IT! *Boat trips take two hours (each way) from Caye Caulker, Ambergris Caye or Placencia. Alternatively, you can stay at a dive resort on a nearby cay.*

214

Stroll around an urban rainforest in Stanley Park

CANADA // Enter Stanley Park, where the gentle swish of dense forest gives way to rippling ocean backed by snow-capped mountains, and you'll swear you're in the Canadian wild, not steps from Vancouver's glassy skyscrapers. The 400-hectare (1000-acre) woodland is one of North America's largest urban green spaces, and wafts a mystical natural aura. Easy trails take you into the thick of it, especially the seawall path that rings the park. Amble past colourful totem poles, swooping blue herons and harbour seals that poke their heads out of the waves just offshore. Log-strewn Third Beach puts up pyrotechnic sunsets. Wherever you are in the expanse, part of the beauty is knowing you're never far from a craft-beer pouring pub.

SEE IT! *Bus 19 stops by the park's Georgia St entrance. Visitors pack the seawall in summer. Arrive early morning or early evening for more tranquil times.*

Witness high, dry drama in volcanic Teide National Park

SPAIN // Spain's most popular national park isn't actually in Europe – it's located 320km (200 miles) off the coast of North Africa on the arid island of Tenerife. Here lies the country's highest mountain, 3718m (12,198ft) El Teide, whose massive volcanic hump dominates the whole island and whose brooding silhouette casts the world's largest shadow over the surrounding sea. The park's dramatic lunar landscapes attract around four million annual tourists. Most take a cable-car up the mountain leaving the abundant trails refreshingly light on foot traffic.

SEE IT! ***Tenerife has two international airports. A daily bus runs to and from the park from the town of Puerto de la Cruz.***

Enjoy spontaneous street theatre in Oaxaca's Zócalo

MEXICO // The zócalo (plaza or square) is the hub of life in any Mexican settlement, not least cosmopolitan Oaxaca, a city well known for its vivid manifestations of indigenous culture. For a colourful dash of local theatre, grab a seat in an alfresco cafe under the arched *portales* (arcades), and cast an eye over the mariachi bands, hard-working shoe-shiners, dancing seniors, wise-cracking balloon sellers and anyone else with a story to tell. You'll never feel so alive.

SEE IT! ***Oaxaca has an international airport, and excellent bus connections to cities all over Mexico. The Zócalo is slap-bang in the pedestrian-friendly city centre.***

Circumambulate Mt Kailash, most sacred of mountains

CHINA // With its sheer black-rock summit, Mt Kailash (6638m/21,778ft) could be the world's most venerated holy place. Hindus believe it's the home of Lord Shiva, and myths speak of it as the birthplace of the whole world. For such a holy heavyweight, it draws relatively few pilgrims because of its inaccessibility, although upgrades to tourist infrastructure could change that. Once you've marvelled at its awesome aura, complete a *kora* (circumambulation) around the base – an unforgettable three-day journey across a stark, cold landscape.

SEE IT! ***Mt Kailash is accessed via the small town of Darchen, starting point of the*** kora ***and a lonely 1200km (745 miles) from Lhasa.***

© Fotografie-Kuhlmann / Shutterstock

An elephant on the road during a sunset game drive in Etosha National Park

218

See wildlife in abundance at Etosha National Park

NAMIBIA // Etosha offers animal spotting made easy, with its wealth of wildlife practically coming to you. The park is dominated by an immense saline pan that glitters brilliant-white under the Namibian sun, and gives the park its name ('Great White Place' in the local Wambo language). When the rains fall, it turns for a few days into a shallow lagoon teeming with pelicans and up to a million flamingos. Rimming the pan to the south is a string of waterholes, which attract an almost constant parade of wildlife during the dry season.

Self-drive safaris are the norm in Etosha, which benefits from an excellent network of well-maintained roads. All you have to do is find a spot by one of the precious ponds and wait for the elephants, lions, rhinos, zebra, oryx and other 100-odd mammal species to come by. And the action doesn't stop after dark. Rest camps within the park have floodlit waterholes, where you can sit and watch as one animal after another comes to quench its thirst.

SEE IT! *Etosha is a six-hour drive from Windhoek. Visit during the dry winter months from June to October, when animals are at waterholes and grass is low.*

Opposite, from top: the chalky rock face of Seven Sisters Cliffs; Cliff Palace, Mesa Verde National Park

Stand tall above England's whitest cliffs, Seven Sisters Chalk Cliffs

ENGLAND // Flagstaff Bottom. Rough Brow. Brass Point. Beachy Head. Each undulation on the epic Seven Sisters has a name, many of them rich with poetry. And wouldn't you want a name to inspire you if you sat high above the English Channel, the South Downs at your back and a great chalk cliff beneath you? The views are, unsurprisingly, wonderful – indeed, the Seven Sisters are often the screen double for the more famous but grubbier White Cliffs of Dover, since the crash and tear of the sea beneath keeps the cliff faces a brilliant white.

Many come, snap that photo and leave, but there's plenty to do here, with a beach at Birling Gap, walks along the South Downs Way (which runs along the top of the cliffs) and even a lighthouse you can stay in. Nearby are handsome villages and fine country pubs.

☛ SEE IT! *Buses run from Eastbourne and Brighton to Seven Sisters Country Park, from where the cliffs are a short uphill walk. There are several car parks.*

Climb to mysterious cliff dwellings at Mesa Verde National Park

USA // More than 700 years after its inhabitants disappeared, Mesa Verde retains an air of eeriness. No one knows for sure why the Ancestral Puebloans left their elaborate cliff homes in southwest Colorado. The sandstone structures are incredibly well preserved, thanks to the dry high-desert air. And they're wild to explore. You don't just walk into the buildings of Mesa Verde. You clamber up lofty ladders set against sheer cliffs to reach the abodes, just like the ancients did. Then you crawl and climb some more as you take in the kivas, plazas, stone towers and painted murals. It's all quite transporting: the elevation puts you next to the mammoth sky, and every breath of whistling wind feels as though it carries the secrets of the departed.

☛ SEE IT! *The nearby towns of Cortez and Mancos make good bases for exploring Mesa Verde. Mid-May to September is the best time to go.*

219

220

The vast wilderness of the Dana Biosphere Reserve, home to 215 bird species, including kestrels

Trek the blockbuster landscapes of Jordan's Dana Biosphere Reserve

JORDAN // Only in a country so packed with big-ticket treats could somewhere like Dana Biosphere Reserve be so overlooked. It's the largest nature reserve in Jordan, fringed by vast sandstone escarpments that lead down into shady green valleys rich in bird and plant life. Sightings of ibexes and gazelles aren't uncommon.

The reserve is one of the linchpins of the epic Jordan Trail hiking route that runs the length of the country. The five-day trek from the pretty Ottoman-era village of Dana along the back route to Petra is regularly hailed as one of the finest in the world, particularly in spring when the route is speckled with wildflowers. The trek takes you through a variety of ecosystems, past ancient copper mines, modern Bedouin camps, and the award-winning Feynan Ecolodge, before threading a series of dramatic gorges to emerge behind the rose-pink facades of Petra.

For those shorter on time or energy, the reserve has trails mapped out to explore in a short morning or long day, and superb guides to show the reserve's riches at their best.

SEE IT! ***The entrance is off the Kings' Highway, near Tafila. Private taxis can be organised from Petra.***

222

Bear witness to an agonising history on Île de Gorée

SENEGAL // An eerie calm shrouds Gorée. Bougainvillea-flushed lanes and bright colonial buildings fill the sand-swept, car-free island. But the stillness is not so much peaceful as meditative, as Gorée's structures bore witness to the Atlantic slave trade. The House of Slaves is the era's chilling memorial; the Door of No Return opens to the sea. There's debate about how many victims passed through, but no one disputes the stark reminder it evokes.

SEE IT! *Gorée Island is 3.5km (2 miles) off the coast from Dakar. Ferries depart every one to two hours for the 20-minute journey.*

223

Visit Korea's DMZ, the world's most fortified border

NORTH & SOUTH KOREA // To set foot in totalitarian North Korea without going there, visit Korea's DMZ (Demilitarised Zone). This heavily guarded, landmine-lined buffer, 240km (150 miles) long and 4km (2.5 miles) wide, cuts through the peninsula, dividing it into two countries. Join a tour from Seoul to the Joint Security Area, where blue office buildings straddle the ceasefire line, guarded by South Korean forces and the UN Command. Inside, you cross momentarily into North Korea's territory, where intimidating soldiers peer through the glass at you.

SEE IT! *Recommended tour operators are Panmunjom Travel Center and USO (United Services Organizations). A set of 'peace trail' hikes launched in 2019 to help ease tension and offer visitors a new view.*

224

Roam bonnie banks and deep waters at Loch Lomond

SCOTLAND // Britain's biggest lake is 37km (23 miles) long and 8km (5 miles) wide. That might not make Loch Lomond a global giant, but this 'bonnie, bonnie' place, celebrated in song and bridging the Lowlands and the Highlands, looms large. Its southern stretches are spotted with islands; in the north the waters deepen and mountains such as giant Ben Lomond gather. There are cycling trails all around its historic shores, and there's excellent walking on the eastern shore.

SEE IT! *Loch Lomond is easily accessed from Glasgow. The western shore can get busy, with the eastern shore usually quieter.*

222

© Frans Lemmens / Getty Images

Climb the dunes of Erg Chigaga

MOROCCO // Climbing to the top of a 300m-high (984ft) sand dune is no easy task. Your feet slip back again and again and every step, conspiring with the audacious heat of the desert, feels like three. But trust us, it's worth it: nothing beats gazing out over the largest sand sea in Morocco and then spending the night in the Sahara, listening to the wind whipping up small sandstorms and create shingle avalanches off the dunes. Erg Chebbi may be better known, but Erg Chigaga steals hearts because of its sense of isolation. Small semi-permanent camps hide in its 40km (25 miles) of troughs, making the experience quiet and enveloping, with some of the best Milky Way viewing you'll ever find.

SEE IT! ***Erg Chigaga is 56km (35 miles) southwest of the town of M'hamid; travel by camel for a five-day round trip, or hire a 4WD.***

© Anton_Ivanov / Shutterstock

Dwell on surrealism at Museo Frida Kahlo

MEXICO // If you want to delve into the film-worthy life of surrealist Mexican artist Frida Kahlo, it's near-obligatory to visit the Casa Azul in Coyocán, a village-like suburb of Mexico City. This is where Frida was born in 1907, died in 1954 and lived a good part of her tumultuous life with the celebrated muralist Diego Rivera. Painted a vivid shade of blue, the house is scattered with fascinating artefacts, including jewellery, Mexican crafts, portraits reflecting Frida's penchant for leftist politics and a good cross-section of her own sometimes disturbing work. The place has become something of a pilgrimage site for Kahlo-philes since the artist's dramatic rebirth as a feminist icon in the 1990s.

SEE IT! ***From the city centre, take Metro line 3 to Coyocán and walk the last 1.5km (1 mile).***

227

Ponder the mysteries of Stonehenge

ENGLAND // Beholding the familiar outline of Stonehenge erupting from Salisbury Plain is mind-bending – especially through a car windscreen from the A303. Ancient and modern Britain collide here, and it's gloriously confusing. How did the sarsen slabs and great bluestones get here, some all the way from Wales, 5000 years ago? Why are they arranged thus – is it a calendar, a cemetery, a place of healing? And how the hell did Neolithic people balance them like vast Jenga blocks?

The quest to control and define such mysteries has persisted through the ages: 12th-century historian Geoffrey of Monmouth connected them with giants, New Age travellers claimed them for the people, and planners are currently seeking to build a road tunnel beneath them, to the dismay of archaeologists eager to find new structures in the soil. Find your own Stonehenge by exploring the site and visitor centre.

SEE IT! *Buses run from Salisbury. Direct access to the stones is usually via (heavily subscribed) Stone Circle Access tours.*

© Andy Roland / Getty Images

228

Time-travel to medieval marvel Tallinn Old Town

ESTONIA // Picking your way along the narrow, cobbled streets of Tallinn's impossibly picturesque Old Town is like strolling into the 15th century – not least because local businesses tend to dress up their staff in peasant garb. But you'll forgive the twee theatrics as you walk past old merchant houses, find hidden medieval courtyards and ascend looming spires to discover sweeping views over the city. The Town Hall Square, Toompea Castle and the onion-domed Alexander Nevsky Orthodox Cathedral compete for photographic attention. Tallinn Old Town's near-complete city walls secured its Unesco recognition, but it was its fairy-tale charm and two-tiered setting that made it a tourist magnet. Though it's just about coping with the burden, visiting in winter, when the rooftops are laden with snow, will deliver maximum romance with far fewer crowds.

SEE IT! *Tallinn is the capital of Estonia. In summer, cruise-ship tourists mob the streets of the Old Town, but they leave by 5pm.*

© kavalenkava / Shutterstock

© Michael Zech Fotografie / Shutterstock

Wander through bygone centuries on Mozambique Island

MOZAMBIQUE // Tiny in size, powerful in presence, this island – for centuries, the capital of Portuguese East Africa – has enchanted all who have stepped ashore since Vasco da Gama landed here in 1498. The European governors and Arab traders may have gone, but relics of the past loom large, including the 16th-century Fort of São Sebastião, and the Palace and Chapel of São Paulo. A lively fishing community still calls the island home; see their colourful nets drying on small beaches, stroll across once-grand praças, and take in the rich mix of influences.

SEE IT! *Nampula, 180km (112 miles) west, is the gateway, with regular public transport. Stay at a small hotel and explore on foot.*

Go back to nature in the Swiss National Park

SWITZERLAND // Nature is left to its own thrilling devices in the remote and ruggedly mountainous Swiss National Park. The first national park to be established in the Alps – and still the only one in Switzerland – it stays true to its original conservation ethos: no trees are felled, no meadows cut, no animals hunted. The result is a spirit-lifting, Eden-like space of larch forests, jagged peaks and lake-lapped meadows that is home to ibex, chamois, marmots and recently reintroduced bearded vultures.

SEE IT! *The park is in Graubünden in Switzerland's southeast. Zernez is the gateway and home to the Swiss National Park Centre.*

Stand face to face with moai at Easter Island's Ahu Tongariki

CHILE // This is the reason you came all the way to Easter Island, flying hours over open ocean to land on this bare triangle of earth. To gaze upon the *moai* – the mammoth, chiselled stone heads, set on platforms called *ahu*. Ahu Tongariki is the largest, with 15 *moai*, including the island's largest, weighing in at 86 tonnes. They were all toppled during the 18th and 19th centuries, with some 50 re-erected and restored. Stand before them as they stare inscrutably into the distance and marvel at the enigma of history.

SEE IT! *There's one six-hour flight daily from Santiago to Easter Island. Ahu Tongariki is in the northeast.*

Bruce Poon Tip's Top Five Places

Canadian entrepreneur and philanthropist Bruce Poon Tip is the founder of award-winning small group adventure travel company and social enterprise G Adventures.

01

WINDHOEK, NAMIBIA – I love deserts and I love traditional African safaris. Plus Windhoek is one of the most fascinating settlements in the middle of Africa. I still hope to live there one day.

02

BELIZE – The ultimate place to get lost: a country that has only 370,000 people is *the* place for anonymity and freedom. It's also where I created the very first G Adventures brochure, in a beach village called Hopkins.

03

TIBET – Traveling to Tibet forever changed my world view, and showed me the connections between life, business, heart and spirituality. This is where I learnt to make decisions with my heart, not my mind.

04

BATH, UK – On a sunny day Bath is one of the most charming English towns. I've always enjoyed its history, architecture and antiques. It's got a lovely quietness that is pure Britain.

05

GALÁPAGOS – The Galápagos is an incredible place that mixes isolation and beauty with education and an understanding of Darwin's theory of evolution. As I've got older I really appreciate more remote spaces.

232

Get to the heart of the Guatemalan Highlands on Lago de Atitlán

© Lucy Brown - loca4motion / Getty Images

GUATEMALA // Ringed by picture-postcard volcanoes, its shoreline dotted with Maya villages, this vast lake serves as an irresistible invitation to kick back and soak up the serenity. With plenty of Spanish schools, bars and good accommodation, San Pedro La Laguna is a popular town in which to bed down for a while. Hike the San Pedro volcano, kayak or paddleboard between villages, take a yoga or cooking class on the lakeshore, or simply relax and enjoy the beautiful views with a freshly brewed espresso – it's one of Guatemala's premier coffee regions, after all.

SEE IT! ***The lake's gateway town of Panajachel is 2½ hours from Antigua by shuttle bus.***

© Justin Foulkes / Lonely Planet

© Gregory Gould / 500px

Be awestruck by age-old architecture at Taos Pueblo

USA // One of the oldest continuously inhabited communities in the USA, Taos Pueblo is renowned for its multi-storey adobe complexes built nearly 1000 years ago. The honey-coloured buildings, set against the backdrop of the soaring Sangre de Cristo mountains, look much as they did when the first Spanish explorers arrived and believed the settlement to be one of the fabled Seven Cities of Gold. Today, around 150 tribal members live on site and provide tours revealing Pueblo Indian history and culture.

☛ SEE IT! *Taos Pueblo is 5km (3 miles) from Taos, New Mexico. The Pueblo closes February to mid-April, and at other times for ceremonies and events.*

Explore ancient rainforests and First Nations culture at Haida Gwaii

CANADA // To walk among the totem poles of Haida Gwaii is to enter a lost world. It's a place where bears roam and bald eagles soar, where sea lions, whales and orcas frolic offshore, and old-growth rainforests feed some of the world's largest spruce and cedars. The ancient Haida culture resonates powerfully across this sparsely populated archipelago off British Columbia's northern coast. It's on particular display in Gwaii Haanas National Park, the islands' number-one attraction, where local guides will take you by kayak or inflatable boat to age-old villages.

☛ SEE IT! *Ferries between Prince Rupert on the mainland and Graham Island take around seven hours. Plan well in advance, especially during the July-to-mid-August peak season.*

© Niyazi Uğur Genca / 500px

© Kris Davidson / Lonely Planet

235

Stroll through the Greco-Roman city of Ephesus

TURKEY // Ephesus is one of the most phenomenal ancient sites in the world. It was once an important city for the Greeks, and the Romans made this great port the capital of Asia Minor. The Temple of Artemis, the biggest ever built, is one of the Seven Wonders of the Ancient World.

Most of the ancient city has yet to be excavated, which makes you marvel at what still lies beneath. What's exposed is wonderfully impressive – from the towering columns of the Library of Celsus to the grand amphitheatre. The mosaics, frescoes and gladiator graffiti of the seven terraced houses show how rich Romans lived. As you walk down the 210m-long (690ft), marble-paved Curetes Way, imagine the other sightseers in togas to evoke this great classical city.

SEE IT! *Nearby Selçuk is the best base – arrive early or late in the day to avoid the heat and the worst of the crowds.*

236

Submit to the neon-lit hedonism of the Vegas Strip

USA // That thing that happens when you take the ideals of freedom and abundance to their extremes? Right here, in this 6.7km (4.2-mile) section of Las Vegas Blvd – aka the Strip – where one whopping casino resort after the other rises up, each trying to outdo the next with dazzling sights. As you glide down the street see fountains that dance to Italian opera, an erupting volcano, the Eiffel Tower, an Egyptian pyramid, even the Statue of Liberty, and rub your eyes in disbelief. Things get more outrageous behind the walls: white tigers prowl, sharks swim, Elvis marries couples non-stop, and Lady Gaga puts on a show. Meanwhile, vast fortune awaits, if only you win big at poker, blackjack or the slot machines.

SEE IT! *Rideshare vehicles are best to get around the Strip. Go April–June or September–November for optimal climate and crowds.*

© Attila Jandi / Shutterstock

© Pecold / Shutterstock

Hang out with rhinos at Mkhaya Game Reserve

ESWATINI // To spend time in this highly protected reserve, whether led on a walking safari or guided in an open-topped 4WD, is to be humbled by the sheer size and majesty of the rhino. There's perhaps no better place on the planet to have encounters with both the black and white species in the wild. Listen to them breathe and watch them do everything, from scratching an itch to taking a dip to beat the midday sun. While on the move between rhino sightings you'll also learn about the reserve's flora and fauna, and have the opportunity to observe the other wildlife species that Mkhaya protects, such as the sable, Livingstone's eland, crocodile, hippo, buffalo, tsessebe (antelope), impala, giraffe and serval.

SEE IT! ***Visits must be pre-arranged, and there are no self-drive options. It's best to spend at least one night.***

Step back in time at Stone Age Skara Brae

SCOTLAND // Before the Pharaohs ruled (let alone dreamt of pyramids), with Stonehenge yet to be cleaved from rock, the village of Skara Brae was already a busy Stone Age centre. Buried in coastal dunes for centuries until an 1850 storm exposed the houses underneath, this Orkney site is so well preserved it can feel as though the inhabitants have just slipped out to go fishing. Even the stone furniture – beds, boxes and dressers – has survived the 5000 years since a community lived and breathed here. Start with the excellent interactive exhibit, move on to a reconstructed house, and then finish up at the excavation itself. This is the best window into Stone Age life you'll ever see.

SEE IT! ***Skara Brae is 13km (8 miles) north of Stromness on Mainland, Orkney's biggest island. Catch the ferry from Scrabster.***

Yellow baboons, zebras and guinea fowl share a waterhole in Majete Wildlife Reserve

See Malawi's Big Five in Majete Wildlife Reserve

MALAWI // Majete is both a conservation success story and a game park that is putting Malawi on the safari map. When NGO African Parks took over the reserve in 2003, it was a poached-out wasteland with just a few antelopes left in its *miombo* woods. Today, it's the only Malawian reserve with the full complement of the Big Five (lion, leopard, elephant, rhino and buffalo).

The transformation has taken 15 years of hard graft in the bush. African Parks has improved infrastructure, worked with local communities and reintroduced almost 3000 animals of 15 species, increasing the wildlife population to well over 12,000. Four cheetahs were recently translocated to the reserve from South Africa, marking the big cats' first appearance here in 20 years. For visitors, this adds up to a still-unspoilt reserve with a clutch of great-value bush lodges, staffed by the friendly folk of the 'warm heart of Africa'.

SEE IT! ***Majete is 70km (43 miles) southwest of Blantyre. Just inside the main gate, meet locals and witness Majete's social benefits in the interpretative centre, restaurant and campsite.***

© Matyas Rehak / Shutterstock

A marsh deer wading through the water in Parque Nacional Iberá

Discover wildlife staging a comeback in Parque Nacional Iberá

ARGENTINA // With the creation of Iberá National Park in 2018 alongside a large provincial reserve, one of the continent's principal freshwater wetlands also became one of Argentina's largest protected areas. Think South American safari: this watery expanse has become a hub for rewilding native species in the Southern Cone. The wildlife watching is thrilling. Camp among docile capybaras, kayak through lily-strewn waters to spy prehistoric-looking caimans, and fix your binoculars on herds of rheas and flocks of green-tipped macaws. Native jaguars and giant river otters are coming back, thanks to the work of Tompkins Conservation, based in the island outpost of San Alonso.

This corner of Argentina has a proud *criollo* (Spanish and indigenous) culture. Take in the region's gaucho routes with a semi-aquatic horseback riding tour, and admire the barefoot cowboys, swimming steeds and banner blue skies. Outside Concepción, travel via horse-drawn canoe to traditional settlements on tiny islands in the estuary. Once home to a thousand residents ranching in this watery world, today they have only 30.

SEE IT! ***You can fly to Corrientes from Buenos Aires, or do a side trip from Misiones province by bus if you go to Iguazú Falls.***

241

Ascend the battlements of old Dubrovnik

CROATIA // Renowned as the setting for King's Landing in *Game of Thrones*, these sublime city walls sit on a headland jutting out into the blue waves of the Adriatic, their mighty towers rising over masts of anchored ships with green Dalmatian islands as a sublime backdrop. The ramparts standing today were built between the 12th and 17th centuries and have never once been breached – though many have tried. The biggest threat today is not invaders, but over-tourism.

SEE IT! ***For the fewest crowds, visit before the daily armada of cruise vessels unload from Dubrovnik's harbour.***

242

Buckle up for a wild road trip on Norway's Trollstigen

NORWAY // A brake-screeching, gear-crunching monster of a mountain road, the 55km (34-mile) Trollstigen (Troll's Path) wiggles precariously around 11 hairpin bends. It's shockingly steep, narrow and serpentine – especially when you have to reverse for oncoming traffic – as if a troll has cleft the earth to reveal ragged peaks and thrust down a thin ribbon of road almost as a joke. The highest point is 858m (2815ft) Stigrøra, with sensational views of the surrounding wilderness, close to the crash-bang spectacle of Stigfossen falls.

SEE IT! ***The Trollstigen Pass links the town of Åndalsnes with the fjord-side village of Valldal, mid-May to October (snow permitting).***

243

Marvel at Mexico's National Museum of Anthropology

MEXICO // The ultimate in Mexican museums resides in a suitably monumental building in Mexico City, pulling together a comprehensive greatest hits of pre-Columbian culture, from the pioneering Olmecs and their colossal head sculptures to the cosmology-obsessed Aztecs and their famous 'sunstone'. Anchored by a pond-filled courtyard that's part-covered by a concrete umbrella, the modernist building utilises indoor and outdoor space to help display its colossal archive.

SEE IT! ***Free one-hour guided tours in English are worthwhile for making sense of Mexico's complicated history.***

Danny MacAskill's Top Five Places

Scottish mountain biker Danny MacAskill is a Red Bull athlete and the star of over 300 million video views on YouTube. He uploaded his first tricks over ten years ago from the streets of Edinburgh, paving the way for today's extreme sports vloggers.

MARITIME MUSEUM, SAN FRANCISCO – It was here I learned about Kenichi Horie, who traversed solo from Japan to San Francisco in the 1960s in a boat the size of a closet.

GIANT ANGUS MACASKILL MUSEUM, ISLE OF SKYE – This little thatched building is dedicated to the life of Angus MacAskill, who was the tallest man (7ft 8in/2.3m) to live without gigantism, and travelled the world.

OLD SCHOOL RESTAURANT, DUNVEGAN, SCOTLAND – While you're visiting the Giant Angus MacAskill museum, this is the best place to get some traditional Scottish food. Lots of juicy prawns, scallops and black pudding.

VILLA EPECUÉN, ARGENTINA – I made one of my many trials videos here, but it's better-known for being wiped off the map after it was engulfed by flood water from Lake Epecuén in the 1980s. As the water slowly receded it left this incredible, kind of apocalyptic landscape.

ALPS EPIC TRAIL, SWITZERLAND – If there's one trail that anyone with a mountain bike needs to ride in their lifetime it's this. The first section of the all-mountain trail is one of the most thrilling, technical and fast downhill stretches that I've ever done. It's the most fun I've had on a trail.

Cruise the island-studded spectacle of Halong Bay

VIETNAM // Halong translates as 'where the dragon descends into the sea', and legend claims the islands of Halong Bay were created by a great dragon whose flailing tail gouged out valleys and crevasses. When it finally plunged into the sea, the area filled with water, leaving only the pinnacles visible – 2000 or more islands rising from the emerald waters of the Gulf of Tonkin. This landscape is often compared to Guilin in China or Krabi in Thailand, but Halong Bay is more spectacular. Its immense number of limestone islands are dotted with wind- and wave-eroded grottoes, and their sparsely forested slopes ring with birdsong. You won't have it all to yourself, but consider cruises focused on Lan Ha Bay, near Cat Ba Island, for less-crowded karst views.

SEE IT! ***Opt for an overnight cruise, sleeping on the bay. November's sunshine and fewer crowds make it a good time to visit.***

© Aureliy / Shutterstock

© Justin Foulkes / Lonely Planet

245

Hike up Central Asia's highest peaks in the Tian Shan

KAZAKHSTAN // The giant Tian Shan mountain range sweeps up over from the Himalaya into Central Asia and down to meet the Pamirs, ducking into the eastern corner of Kazakhstan on its way. A country the size of western Europe, Kazakhstan is chock full of incredible landscapes, but the Tian Shan is a perennial favourite. The country's old capital, Almaty, is nestled up against the range's lower slopes, and not far away are skiing, trekking and hiking options galore. Real mountaineers might want to scale 7010m (22,998ft) Mt Khan Tengri, one of the range's most beautiful and treacherous peaks, which sits on a corner where Kazakhstan meets China and Kyrgyzstan.

SEE IT! *A day trip from Almaty to hike near turquoise Big Almaty Lake, surrounded by Tian Shan peaks, is an accessible way of enjoying the mountains.*

246

Cycle through the architectural legacy of Ayuthaya

THAILAND // Alas, despite it having been the capital of Siam between 1350 and 1767, the ruins of Ayuthaya are unfairly considered but a poor relation to other Southeast Asian historic sites, such as Angkor. Ayuthaya can't compete on size or scale with its Cambodian counterpart, but it still offers a tantalising glimpse into glories past. The core of the ancient city is on Ayuthaya's river island, where more than a dozen partially restored ruins can be found, along with several working temples. At its zenith, 400 sparkling temples stood here, before the city was devastated by an invading Burmese army. There's also much pleasure to be had in cycling past minor ruins in the green expanse of Somdet Phra Sri Nakharin Park in the southwest of the island.

SEE IT! *Ayuthaya lies an hour north of Bangkok and is connected by boat, bus and train. Hire a bicycle to explore the site fully.*

Turquoise waters of the Llanganuco lakes in the Cordillera Blanca mountains

Ascend into South America's hiking heaven in the Cordillera Blanca

PERU // The Andes mountain range is the planet's longest, rucking up a raft of ridges over 5000m (16,400ft) across seven South American countries, and yet Peru's Cordillera Blanca caps almost all of it, not for height (that's Argentina's Aconcagua) but for the sheer, sustained high-altitude beauty and terrific trekking opportunities. The mountains festooning this region are rightly called *nevados* (snowcaps): high enough, with 18 summits cresting 6000m (19,685ft), to be under more or less permanent snow cover. But it is the montage of cerulean lakes, pearly glaciers, stark *quebradas* (ravines) and gushing waterfalls below the mountains that most impresses. Much of the best scenery is threaded through by the popular, four-day Santa Cruz trek from Cashapampa, but the region is replete with superb mountain hikes.

SEE IT! ***Huaraz is the base for most outdoor activities hereabouts, although on a multi-day trek like the Santa Cruz you will be camping in the mountains.***

© matthi / Shutterstock

248

Crack Edinburgh's storied spine on the Royal Mile

SCOTLAND // With Edinburgh Castle at one end, Scottish Parliament at the other and St Giles Cathedral in the middle, the Royal Mile embodies history like nowhere else in Scotland. It runs up to Castle Rock (the volcanic plug on which the castle sits) like a great, cobblestoned spine, views stretching across to Arthur's Seat and the Georgian New Town, and down to the Firth of Forth. Its sides are dotted with atmospheric alleyways ('wynds'), and beneath its cobbles are medieval catacombs, which you can visit on tours. It throbs with performers, tourists and salespeople hawking shows during Edinburgh's festival seasons. But it's the historic pubs, attractions such as the Camera Obscura and the Scottish Storytelling Centre, shops packed with souvenirs, and some fine restaurants that make this famed street so appealing to all comers.

☛ SEE IT! *The Royal Mile is in central Edinburgh. There's plenty of accommodation, but book ahead at New Year and around August.*

© xuanhuongho / Getty Images

249

Confront the brutality of war at the War Remnants Museum

VIETNAM // Few museums anywhere convey the brutal effects of war as powerfully as the War Remnants Museum in Ho Chi Minh City. For many who only know the 1954–75 war in Vietnam through histories told by the West, this is a rare opportunity to hear the Vietnamese share their own stories about the conflict they call the American War. While some displays are one-sided, there's no denying the photographs: injured bodies, children affected by napalm, atrocities such as the My Lai Massacre – horror and grief up close. Among the many other exhibits are contraptions of cruelty, including a replica of the infamous 'tiger cages' that held prisoners on Con Son Island. There is also a section showing images of support for the antiwar movement from around the world – a somewhat positive note.

☛ SEE IT! *This is one of HCMC's most popular sights; visit earlier in the day when it's less busy.*

Step into a magical forest at Arashiyama Bamboo Grove

JAPAN // At Arashiyama's bamboo grove in Kyoto's west, nature wows with effortless simplicity. The tall green stems seem to have duplicated endlessly in every direction, and when a breeze rushes through, it all creaks and rustles – a sound that has been designated one of the '100 soundscapes of Japan'. Walking through the grove is like entering another world and there's an ethereal quality to the light. But go easy with the camera; it's a quality you won't fully capture in pictures.

SEE IT! ***Stay overnight in Arashiyama so you can step out at dawn and enjoy the grove before the rest of humanity arrives.***

Hole up in Cuba's bucolic heart in Valle de Viñales

CUBA // Embellished by soaring pine trees and bulbous mini-mountains that teeter like giant haystacks above placid tobacco plantations, Parque Nacional Viñales is one of Cuba's most magnificent spectacles. This other-worldly karst landscape, pockmarked by primeval *mogotes* (precipitous limestone buttresses), was created by the collapse of a vast prehistoric cave system, and the valleys are still worm-holed with caverns, offering opportunities for caving, rock climbing and trekking.

SEE IT! ***Viñales, connected by bus to Havana, is the portal to the Valle, with abundant*** **casas particulares** ***– family-run homestays.***

Share in the heartbreak at the Museum of Broken Relationships

CROATIA // Reckoning that break-ups are a universal human experience, two Zagreb artists (and former lovers) put out a global call for objects representing the end of relationships. The delightfully eccentric results are now housed in a stark white museum. Unsent love letters. An axe used to chop up an ex's belongings. A wedding dress stuffed in a jar. Breast implants requested by a boyfriend, then removed after the split. Cry, laugh, relate.

SEE IT! ***The museum is at Ćirilometodska 2. It's open from 9am to 10.30pm in summer, and 9pm the rest of the year.***

© Suchitra Poungkoson / Shutterstock

253

Chant with Buddhist monks at Ladakh's impressive Thiksey Monastery

INDIA // Glorious Thiksey Gompa is so huge that, at first glance, it looks more like a village than a monastery. Tumbling over a large rocky outcrop, the layers of traditional white and red buildings house protector chapels, stunning murals and a main assembly hall that features a 14m-high (46ft) Buddha, all spectacularly ringed by arid khaki mountains. Stay overnight at the monastery guesthouse and you can join monks in their morning prayers, a fascinating ceremony that's so popular visitors can sometimes outnumber worshippers. A museum hidden beneath the monastery restaurant displays Tantric artefacts, including a vessel made from a human skull.

SEE IT! *Thiksey is near Leh in the remote north Indian region of Ladakh. From October to May, the only way in is by air.*

© Philip Lee Harvey / Lonely Planet

254

Discover South Luangwa National Park, Zambia's best-kept safari secret

ZAMBIA // Zambia flies under the radar when it comes to name recognition – how else to explain why South Luangwa National Park is not more celebrated as one of Africa's premier safari parks? For scenery, density and variety of wildlife, choice of accommodation and quality of guiding, it's the best park in Zambia, and one of the most majestic in Africa. The varied terrain of dense woodland, oxbow pools and open grassy plains hosts beasts of all shapes and sizes, from massive elephants to lions in abundance. But South Luangwa's greatest attraction is its untamed landscape, where you can immerse yourself fully in the sights, sounds and smells of the bush.

SEE IT! *Access South Luangwa via light aircraft or 4WD. Independent travel is possible, but safari operators smooth the way.*

© Andrew Molinaro / Shutterstock

Gaze at chimpanzees in the forests of Mahale Mountains National Park

TANZANIA // Lake Tanganyika's crystal-clear waters lap coves of powdery white sand that rise almost immediately to forested mountain slopes. Surely this must be one of Africa's most unique and most beautiful national parks, with its contrasts of colour and terrain, unforgettable scenery and sense of peaceful remoteness. In addition, Mahale is one of the world's best places to observe chimpanzees in the wild, with the main group well habituated to humans, thanks to a decades-long primate research project. Expect some strenuous, sweaty hiking with your guide to locate them, and respect the park's strict viewing rules, which have been put in place to protect the chimps. Evenings are a treat, spent relaxing on the beach and watching the sun set in orange- and purple-hued magnificence over the lake.

SEE IT! ***Reach Mahale via boat, road or air from Kigoma. Time with the chimpanzees is limited to one hour per group per day.***

© Christian Declercq / Shutterstock

Tramp the untrammelled Inca Trail to Choquequirao

PERU // Imagine a Machu Picchu – meaning a spectacular Inca citadel perched on a mountain above Peru's jungle – without crowds. This is Choquequirao: a three- to four-day out-and-back hike from the already-remote outpost of Cachora, 60km (37 miles) and a world away from touristy Cuzco. Straddling a broccoli-green ridge at 3000m (9842ft), the ruins mark the refuge of Manca Inca, on the run after he tried to retake Cuzco from the Spanish. The truncated summit, a ceremonial plaza hewn out of the hill above the main complex, and a flight of terraces depicting llamas in stone are the highlights. Despite talk of a cable car to attract the masses, Choquequirao remains an out-of-the-way odyssey. Parts are still unexcavated and a rarely traipsed multi-day trek on from here to Machu Picchu is also possible.

SEE IT! ***Choquequirao is visited via a three-to-four-day hike, either independently or with Cuzco tour agencies.***

© Ryan Carter / Alamy Stock Photo

257

Survey the 360-degree Alps view from the Aiguille du Midi

FRANCE // A jagged finger of rock rising above glaciers, snowfields and icy crags of the mighty Mont Blanc massif, Aiguille du Midi (3842m/12,604ft) is a geographical beacon. The panorama of the French, Swiss and Italian Alps from the summit is breathtaking, especially when peering down through the aptly named Step into the Void's glass floor. Year-round, you can float up in a cable car from the ski-resort town of Chamonix on the vertiginous Téléphérique de l'Aiguille du Midi.

The dizzying views don't end here: between late May and September, you can continue in the Télécabine Panoramique Mont Blanc to Pointe Helbronner (3466m/11,371ft) on the France–Italy border, then another 4km (2.5 miles) in the SkyWay Monte Bianco cable car to Courmayeur, on the Italian side of the mountain.

SEE IT! ***Try to avoid high summer's queues, and dress warmly: the temperature at the top rarely rises above -10°C (14°F).***

© Marco Tomasini / Shutterstock

258

Soak up the haunting atmosphere of Ani

TURKEY // If this medieval Armenian capital and Silk Road *entrepôt* was in western Turkey, it would busier than the Blue Mosque. But its location in a lonely corner of the steppe on Turkey's border with Armenia adds atmosphere to the ruined churches jutting out of the windblown grass, like a scene from *Planet of the Apes*. The distinctive churches with drum cupolas scatter the plain and spill into the river gorge, alongside relics including a broken Silk Road bridge.

Ani's later occupants included the Seljuk Turks, who built the red and black Manuçer Camii in 1072, considered Anatolia's first Turkish mosque. Decline set in under the nomadic Mongols, who cared little when an earthquake devastated the city in 1319. Today, you might be the only person wandering the eerie site among ruins such as Ani Cathedral.

SEE IT! ***Ani is 50km (31 miles) east of Kars. Get there either by taxi or guided tour.***

Cannonball into the beautiful blue hole of To Sua Ocean Trench

SAMOA // This mesmerising blue hole is like a sapphire in a sea of emeralds. In fact, there are two sinkhole-like depressions, dressed in dripping tropical greenery and sparkling with water that flows through an undersea tunnel from the ocean. After clambering 20-odd metres (65ft) down a steep wooden ladder, you can swim under a broad arch of rock to the hidden second pool.

SEE IT! *To Sua is tucked away on the southeast coast of Upolo; look out for the faded sign on the Main South Coast Rd, near Lotofaga.*

Dig your own spa at Hot Water Beach

NEW ZEALAND // On the eastern side of the Coromandel Peninsula is a natural phenomenon unlike any other. For two hours either side of low tide, natural hot-spring waters well up beneath the sand. Visitors dig in to create hot tubs, relaxing in the warm sandy water as the surf breaks just feet away. Nearby are the ruins of one of the earliest villages of the Māori Ngāti Hei tribe, and the golden sands of Cathedral Cove.

SEE IT! *The beach is a 3½-hour drive from Auckland; local shops rent shovels.*

Zip to treetop adventures on the Gibbon Experience

LAOS // Sure, there are other zip lines in the world, and other wildlife-filled national parks. There are even other places where you can spend the night in the trees. But the Gibbon Experience combines all of these into one unforgettable jungle adventure. Trek through Nam Kan National Park, buzz across the valleys on a series of zip lines, spot wildlife, sleep, eat and shower in treehouses some 40m (130ft) up, then wake to the eerie whistling of the elusive gibbons' morning song. What's more, tourist dollars here go towards supporting local communities and conservation projects.

SEE IT! *The Gibbon Experience office is located in Huay Xai. Take a flight from Vientiane or, even better, the two-day slow boat from Luang Prabang.*

262

The peerless pools at Semuc Champey, fed by the Cahabón River

Cool off in the jungle-shrouded swimming hole of Semuc Champey

GUATEMALA // What sounds more refreshing in the tropics than a dip in a shady pool in the middle of the jungle? How about taking a dip in a series of perfectly stepped pools linked by a series of natural rock bridges, and with colours that glitter from turquoise to emerald? This is what Semuc Champey offers, at the end of a rough rainforest road, the bumps of which make the water seem even sweeter. Dive in and you'll quickly understand why many Guatemalans consider Semuc Champey to be the most beautiful spot in the country.

SEE IT! *Semuc Champey is close to the travellers' hub of Lanquín, from where there are local pick-ups to the pools, as well as dedicated shuttle services from hotels and hostels.*

© Gaspar Janos / Shutterstock

Haunt the hallways of Transylvania's most storied stronghold

ROMANIA // Bran Castle in Transylvania has every feature a vampire's black heart could desire. It's perched above misty forests, crowned by a fang-sharp tower, and its chambers are peopled by unsettling suits of armour. What it doesn't have is a credible link to either the fictional Dracula – given eternal life by Bram Stoker's Gothic novel – or the historical figure who inspired him, Vlad Tepes ('the Impaler'). The castle was first documented in the 14th century but Tepes, a national hero for his repulsion of Ottoman forces, probably never stayed within its walls. Even if dubbing it 'Dracula's Castle' was primarily a canny marketing move, Bran Castle's forbidding silhouette and dramatic clifftop setting continue to cast a spell on visitors.

SEE IT! *Half-hourly buses run from Braşov (25km/15 miles southwest). Souvenirs abound, from local honey to tacky T-shirts.*

Let your imagination run riot at Lego® House

DENMARK // Billund's latest big-ticket attraction (opened in 2017) is itself designed to look as though it's made from giant plastic bricks. Inside you'll learn all about the art of 'playing well' – the word 'Lego®' is a portmanteau of the phrase in Danish: *leg godt*. A museum recounts the history of the company, and temporary exhibitions, such as a *Star Wars* retrospective, will delight, with retro sets on display from the archives. Impressive, too, is the Tree of Creativity, a 15m-tall (50ft) structure made with over six million bricks. From the moment you arrive, you'll be itching to get your hands on some plastic pieces to build your very own creations, and there's ample opportunity: themed areas allow you to channel your inner child in creative and thoughtful ways.

SEE IT! *Book ahead on the website for cheaper tickets. Billund is served by an international airport but, as yet, there's no train station.*

Hazen Audel's Top Five Places

Hazen Audel is a TV presenter, biologist and natural history guide. His series Primal Survivor *documents his adventures living and working alongside indigenous people in remote parts of the world.*

01

ECUADOR – This is my second home and where it all started for me. It's relatively small but has great cultural and geographic diversity – you've got the highlands, the coast, the rainforest and even desert areas.

02

NORTHERN CANADA – You have to get used to the cold but what spectacular beauty. I've spent time with the Cree in the Hudson Bay area: it's amazing that people can thrive in these hostile environments, and the wildlife is astonishing – wolves, bears, walrus, caribou.

03

MONGOLIA – I lived there for about three weeks with a local tribe and it was the most fascinating place I've ever been. They still hunt with eagles. It's a harsh environment and you have to be completely self-sustaining.

04

GREECE – When I think about where I might live in later life it would be Greece. Half of my heritage is Greek, and I love their priorities in their relationships with family and community, and their sense of island life.

05

THE SOLOMON ISLANDS – Its reefs are in wonderful shape and the snorkelling is fantastic. I studied tropical botany in Hawaii, so I love that Polynesian, South Pacific sort of rainforest. The plants here are unique and there are so many tiny biospheres to explore.

Explore the mystical prehistoric tombs of Brú na Bóinne

IRELAND // If you're talking about significant mounds of rock, England's Stonehenge comes with a certain historical fame, but Ireland's Brú na Bóinne has the clout of a thousand extra years. It's an extraordinary sight and a testament to the mind-boggling achievements of our prehistoric ancestors: an enormous Neolithic necropolis of squat, rotund tombs set in emerald plains.

The pick of the monuments is Newgrange, with its futuristic white stone walls and grass dome above a prehistoric passage tomb. Steeped in Celtic mythology, the purpose of the site is something of a mystery, though it's believed to have been a burial place for kings. At the winter solstice, the rising sun's rays shine through the opening above the tomb's entrance, illuminating the chamber in what is one of the country's most memorable experiences.

☛ SEE IT! *Buses go via Drogheda from Dublin. Visits are by tour only; just 750 tickets are sold on the day, so arrive early.*

Opposite, from top: beach life on the Bacuit Archipelago; look, but down't swim, at Bath's Roman Baths

Dive into an island paradise in the Bacuit Archipelago

PHILIPPINES // Since it has more than 7000 islands, the challenge in the Philippines is which to choose for that perfect mix of sand, palms and coral. We'll make it easy for you: from the dive resort of El Nido in Palawan, you can access a karst limestone wonderland of jungle-capped islets in a perfect turquoise bay. The Bacuit Archipelago has hidden beaches, secret caverns, lost lagoons, archaeological sites and spectacular dive sites. The main form of transport for getting around is the outrigger canoe, though you can also go by kayak or just use your own pair of fins. Take an island-hopping cruise or charter a boat to drop you on an uninhabited island and go *Robinson Crusoe*; either way, you'll discover what ultimate – and increasingly hard to find – Southeast Asian island travel is all about.

SEE IT! ***El Nido is on the southwest Philippines island of Palawan, served by regular flights from Manila.***

Let history wash over you at Bath's Roman Baths

ENGLAND // Baths are special things, and Bath's have been revered for millennia. Among the finds at this gorgeous ancient site – one of the world's best preserved Roman spas – are 12,000 coins and 130 curse tablets. The coins were flung into the hot spring in honour of the goddess Minerva; the Latin tablets wished dreadful punishments on anyone stealing valuables from blissed-out bathers.

The Celts took the waters here, but it was the Romans who built the complex, with its lead-lined great pool, hypocaust and plunge pool. Vapours still rise from the Sacred Spring (a constant 46.5°C), but you can't take a dip here any more. Instead, wander the terraces and admire the statues and digital reconstructions. Then explore the museum, and seek out its two famous heads – a bronze Minerva and a stone gorgon.

SEE IT! ***Bath is a 90-minute train ride from London. The Roman Baths open at 9am (9.30am in winter) and are busiest 11am–3pm.***

266

267

Tour a secret underground city at the Vieng Xai Caves

LAOS // Concealed within the dramatic karst landscapes of Vieng Xai is a complex system of caves that, at one time, sheltered more than 20,000 soldiers and civilians. Open to the public since 2007, the fascinating caves were home to the Pathet Lao communist leadership during the so-called 'Secret War' and US bombing campaign of 1964–73. Basic bunkers were transformed over time into an underground city of hospitals, markets, a printing room, theatre and even a metalwork factory. Fittingly, it takes some effort to get here, but it's worth it.

SEE IT! ***Day trip by taxi or bus from Sam Neua, which is a 15-hour bus ride from Luang Prabang or 20 hours from Vientiane.***

Revel in the biodiversity of Parque Nacional Madidi

BOLIVIA // Nearly half of all mammal species in the Americas live in this 18,000-sq-km (6950-sq-mile) park, a marvel of biodiversity. The park soars from the sweltering tropical rainforest along the Tuichi River to the thin air of the Andes. It has become a model of ecotourism, with initiatives run by indigenous villagers. Ride an outboard canoe deep into the jungle in search of giant otters, jaguars and spectacled bears.

SEE IT! ***Rurrenabaque is the access town, with booking offices for ecolodges and tours. Go from La Paz by propeller plane or day-long bus.***

Gawk at the grisly monastic art of Sedlec Ossuary

CZECH REPUBLIC // The macabre 'bone church' in the basement of former Cistercian monastery Sedlec Abbey would stretch the imagination of the most accomplished gothic horror novelist. Garlands of skulls hang from its vaulted ceiling, bones adorn the altar, and pyramids of bones squat in each of the corner chapels. Who would have thought it was all the work of a local woodcarver, let loose on Sedlec in 1870 with the bones of 40,000 to 70,000 people found in the old abbey crypt? Creepy, yes – but utterly entrancing, nonetheless.

SEE IT! ***Sedlec is in the medieval city of Kutná Hora, an hour east of Prague by train.***

© Anna Minsk / Shutterstock

© Angel Villalba / Getty Images

The astounding Step Pyramid at Saqqara, the world's oldest stone monument

271

Learn how pyramid architecture evolved at Saqqara

EGYPT // Want to discover how the ancient Egyptians figured out how to build a pyramid? Head to Saqqara. Serving as the burial ground for ancient Memphis, this 7km (4-mile) site is best known as home to the Step Pyramid of Zoser. This 60m-high (197ft) pyramid is not only the world's oldest stone monument, but also showcases how chief architect Imhotep developed what would become the most iconic of ancient Egypt's building works.

Beyond the Step Pyramid, Saqqara is a sprawl of funerary complexes with a glut of richly decorated tombs and other pyramids to explore. The Serapeum is where the sacred Apis bulls were mummified and interred; the walls of the Mastaba of Ti are covered with vivid scenes of daily life, acclaimed as the best-preserved examples of Old Kingdom art; and the interiors of the Pyramid of Titi and Pyramid of Unas are painted with cascades of stars. Saqqara has yet to give up all its secrets – several new tombs containing mummified cats were discovered in late 2018.

SEE IT! ***Saqqara is 30km (19 miles) south of downtown Cairo. Start early so you can combine a visit with a trip to Dahshur.***

© Daniel Marian / Shutterstock

Pelicans skim over the waters of the Danube Delta

Float along the waterways of the Danube Delta

ROMANIA // Romania looks almost tropical in the 4187-sq-km (1617-sq-mile) Delta Dunării (the Danube Delta): there are bird migrations of staggering scale, exquisite orchids and giant dragonflies that dart above the water. This wetland of marshes, floating reed islets and sandbars is constantly evolving, its innermost recesses only accessible to small boats. Deservedly inscribed on Unesco's World Heritage list, it's a haven for lovers of wildlife, birdwatchers and fishers who come to immerse themselves in this sanctuary for 45 freshwater fish species and 300 species of bird, including great white pelican and Dalmatian pelican colonies (among Europe's largest). Spotting the impressive roll call of protected species and migrating populations (white pelicans from June to mid-September and red-breasted geese in winter) that flock here is only part of the allure. White-tailed eagles soar overhead while forest glades conceal vividly colourful birds, such as bee-eaters and European rollers. Beautiful, secluded beaches, swathes of marshland and tranquil lakes beg to be explored before you feast on the best seafood in Romania.

SEE IT! ***The Danube Delta is in eastern Romania. Cruise operator Ibis (www.ibis-tours.ro) is mindful of delta conservation.***

© IZZARD / Shutterstock

© niladrilovesphotography / Shutterstock

273

Scamper around a real-live Hobbit town in Alberobello

ITALY // Your imagination will run riot in Alberobello, famed for its kooky, one-of-a-kind architecture. We're talking *trulli* – whitewashed circular dwellings with cute cone-shaped roofs, originally devised to avoid property taxes (it's a long story). Looking like they're straight out a Disney cartoon, these sunbaked houses tumble down the hillsides of Rione Monti and Rione Aia Piccola. You can dine in some, and sleep in others. Was that Snow White? Are you on Earth? Unesco seems to think so, bestowing World Heritage Status on them in 1996.

SEE IT! ***Alberobello is accessible from Bari on the Bari-Taranto train line. From the station, head down Via Mazzini to the town centre.***

274

Ponder mortality at Varanasi's holy Dashashwamedh Ghat

INDIA // It is at Varanasi's smouldering ghats (riverside steps), lining the waters of the Ganges, that India's holiest city is at its most spiritually enlightening. Throughout the day, families bid a very public farewell to their deceased relatives at the burning cremation pyres here; at dawn, lone pilgrims perform silent prayers to the rising sun. At sunset, the city's main *ganga aarti* (river worship ceremony) takes place at Dashashwamedh, making this Varanasi's liveliest and most colourful ghat.

SEE IT! ***Varanasi is in the north Indian state of Uttar Pradesh; overnight trains run from Delhi.***

Discover Yves Saint Laurent's design legacy in Marrakesh

MOROCCO // As the sun dances over the bamboo groves and amplifies the electric blue of the art deco studio, it's plain to see why Yves Saint Laurent fell in love with the Jardin Majorelle. He and his partner Pierre Bergé bought the villa and its gardens to preserve the vision of original owner Jacques Majorelle. Today, while admission queues may be growing (come at 8am if you can), there's little that can diminish the soul-soothing charms of the garden's pools, pavilions and globetrotting flora. Inside, the Musée Berbère beautifully displays Bergé and Saint Laurent's collection of tribal artefacts, including jewellery, textiles and woodwork. Next door, the Musée Yves Saint Laurent is an architecturally daring repository for the designer's haute couture collections.

SEE IT! *The Jardin Majorelle and Musée Yves Saint Laurent are located a 30-minute walk from Djemaa El Fna.*

© Derek Harris Photography / Shutterstock

Marvel at the miracle of Meteora's saintly pillars

GREECE // If you are cutting yourself off from worldly temptations it makes sense to do so somewhere that Mother Nature offers other consolations. So it is with the monasteries of Meteora (variously translated as 'suspended in the air' or 'in the heavens above'), set on giant spears of rock that burst forth from the fertile plains of Thessaly. There were originally 24 of these Greek Orthodox monasteries, built in the 16th century. They were only accessible by rope ladders, so for the novices in residence there was little to do but contemplate the beauty of God's creation. Today, there are only six monasteries open –now with stairs. And, overlooking the surrounding countryside you still get a feel for the rigours – and exaltation – of the monastic life.

SEE IT! *There's nowhere to stay in Meteora itself; make a day trip from nearby Kalambaka, which is fully equipped for travellers.*

© Justin Foulkes / Lonely Planet

© hecke61 / Shutterstock

© Jarno Gonzalez Zarraonandia / Shutterstock

277

Unleash your inner palaeontologist in Drumheller

CANADA // Dinosaur lovers get weak-kneed in Drumheller, a dust-blown town in southern Alberta where the arid terrain preserves bones galore. Palaeontology pride runs high here, and nowhere more than at the Royal Tyrrell Museum. One of the planet's preeminent dinosaur institutions – chock full of fierce-looking skeletons with names such as 'Hellboy' – it's also an active research centre with fossil digs you can join. Kitschy but no less awesome is the world's largest T rex, a 26m (85ft) fibreglass fella who looms over the town like Godzilla and dares you to climb the steps up into his mouth. The Dinosaur Trail extends the theme beyond town on a drive into the surrounding badlands, an eerie landscape of hulking, mushroom-like rock formations that underscores the prehistoric mood.

SEE IT! *Calgary is the gateway to Drumheller, which is a 1½-hour drive northeast. Late June to early September is the prime time to go.*

278

Observe sea turtles coming home in Parque Nacional Tortuguero

COSTA RICA // In this park spanning 11 types of habitat, coastal rainforest takes on the feel of Amazonian jungle – dripping with humidity, clamourous with squawks and squeals, and teeming with life. As your boat floats through lazy lagoons, you'll spot crocodiles sunning on logs, sloths hanging in trees, otters splashing, and howler monkeys making their presence known. And birds – hundreds and hundreds of birds.

Then there's the main attraction: the turtles. Watch a massive mama sea turtle as she hauls herself onto her natal beach, laboriously digging out her nest and dropping in dozens of eggs. Then glimpse hatchlings scurrying to the sea. You can even sign up to help protect these endangered species, besides witnessing this cycle of life, playing out on a dark stretch of beach just as it has for millennia.

SEE IT! *Tortuguero is the best base for exploring the park. Turtle nesting season is March to October, peaking in July and August.*

Admire Piazza San Marco: the drawing room of Europe

ITALY // We've all heard about the crowds, so how did Venice's Piazza San Marco make this list? Quite simply because it is the most beautiful public space in the world, ringed by elegant architecture, furnished with historic cafes, and bookended by a golden basilica. Yes, tourists throng, but the clever trapezoidal space absorbs them without strain. Still, if you really want to experience why San Marco is a must, come at dawn when the sun sets the basilica ablaze, or on cold winter nights when the moonlit grandeur makes you weep at the beauty.

☛ SEE IT! *Take a boat from Marco Polo airport. From the mainland take the train to Venice Santa Lucia, then vaporetto (waterbus) No.1.*

281

© saiko3p / Shutterstock

Wander the windswept D-Day beaches at dawn

FRANCE // Strolling the shoreline of Normandy as the day breaks and the English Channel laps French sand, all seems serene. It's a stark contrast to 6 June 1944, when Operation Neptune sent some 150,000 Allied troops crashing into this coast – landing them on Omaha, Utah, Gold, Sword and Juno Beaches – as the biggest seaborne invasion in history began. D-Day changed the course of WWII; an exploration of the bunkers and a sombre stroll through cemeteries dedicated to the fallen brings home the horrific magnitude of the soldiers' sacrifice.

☛ SEE IT! *The D-Day beaches start 17km (10 miles) northeast of the history-woven town of Bayeux, from where buses serve the coast.*

Embrace the romance of Rajasthan at Jaipur's Amber Fort

INDIA // More palace than castle, picture-perfect Amber Fort is fierce and formidable from the outside, but inside it's all sensuous carvings, delicate inlays and perforated marble screens. Despite the crowds that can sometimes feel like a besieging Rajput army, it's still possible to step through the Suraj Pol (Sun Gate) into the lavish courtyards beyond and imagine yourself slap bang in the middle of the opulent Mughal court.

☛ SEE IT! *Amber Fort is 11km (7 miles) from Jaipur. Skip the elephant ride to the gate; take a few photos with them in the foreground, instead.*

© Sergei Proshchenko / Shutterstock

282

Camp with reindeer herders on the shores of Khövsgöl Nuur

MONGOLIA // It's tough to nab the 'most beautiful' award in a country so rich in natural sites, but Khövsgöl Nuur is Mongolia's star. The country's largest lake, it's deeply sacred to Mongolians and packed with wilderness for visitors. Its northern shores butt up against the Russian border, and there is no question this is a Siberian lake: the water remains frozen for most of the year. During the summer thaw, you can kayak or go boating, and hike and ride horses around its shores, spotting argali sheep, ibex and maybe a wolverine. Added to the natural wonders here are a deep spirituality and the presence of resilient local cultures, including the Tsaatan reindeer herders, who live in small encampments and still practise traditional shamanism.

☛ SEE IT! *There is a settlement at Khatgal, on the lake's southwestern shore, with a few guesthouses. You can camp anywhere 100m (330ft) from the water; the quietest spots are on the eastern shoreline, reachable by 4WD.*

© Kurlin Cafe / Shutterstock

Traitors' Gate at the Tower of London, where many prisoners of the Tudors arrived during the 16th century

See bling in the bloody tower at the Tower of London

ENGLAND // If you manage to keep your head while gazing up at the formidable castle walls, you can see the spot where two of Henry VIII's wives lost theirs. Raised by William the Conqueror to enforce his rule over the unruly English, the Tower of London is not one castle but a succession of them, built on top of and in and out of each other. An astonishing number of pivotal events in English history took place inside its walls – Sir Walter Raleigh was imprisoned here, the princes Edward V and Richard of Shrewsbury were bumped off by Uncle Richard in the Bloody Tower, even notorious gangsters the Kray Twins were temporary residents when the tower was used as a prison in the 1950s. Then there's the Royal Armoury, with King Henry VIII's (probably exaggerated) codpiece and the spectacular Crown Jewels.

SEE IT! *The tower dominates the north bank of the Thames by Tower Bridge; to get your bearings, take a free guided tour led by the cheery Beefeaters (Yeoman Warders).*

284

Stroll through Seoul's regal Changdeokgung palace

SOUTH KOREA // Principles of feng shui dictate the architecture at this Joseon dynasty palace, adopted as the seat of the Korean royal family after the Japanese destroyed the original in the 1590s. Framed by a mountain and a stream, it is a gateway to a 16th-century world of refinement, exemplified by the Biwon (Hidden Garden), where the royals came to write poetry inspired by their surroundings. It's amazing that something so fragile survives amid Seoul's urban sprawl.

SEE IT! *You can reach the palace by subway; ride Line 3 to Anguk station and leave via Exit 3.*

285

Experience idyllic sinkhole swimming at Cenote Azul

MEXICO // On a coastline better known for its all-inclusive resorts, this foliage-surrounded natural site provides a refreshing break from the ubiquitous Riviera Maya hotels where vacationers recline on sun-loungers and play bingo by the swimming pool. Cenote Azul is a limestone sinkhole filled with fresh water where daredevils dive off a 3m-high (10ft) rock into a pool full of fish, grottoes and occasional turtles. Birdlife and iguanas guard the paradisiacal surroundings. Follow a winding path through the jungle and take the plunge.

SEE IT! *The cenote is 30 minutes by bus from Playa del Carmen. There's a small entry fee and snorkelling gear can be rented on site.*

286

Ride the rainbow at Caño Cristales

COLOMBIA // For an enchanted window between July and November this otherwise ordinary river transforms into what looks like run-off from a crayon factory. Blankets of underwater river weed bloom scarlet, the colour contrasting with the yellow sand, black and green river rocks and blue water. In the Sierra de la Macarena National Park, only 200 visitors are allowed daily. Protect the ecosystem by swimming only in designated areas – no sunscreen or bug spray allowed.

SEE IT! *Licensed tour guides are mandatory. Visit via a three-day tour from Bogotá or day trip from the nearby town of La Macarena.*

Below: the National Mall and Washington Monument. Bottom: the Lincoln Memorial at sunrise

Get chills at the US capital's monumental National Mall

© Albert Pego / Shutterstock

USA // The Mall is often called 'America's Front Yard', and it is indeed a lawn, unfurling 3km (1.9 miles) of scrubby grass in the heart of Washington, DC. It just happens that the nation's most notable monuments and hallowed marble buildings dot it. At the east end, the domed US Capitol rises up, the government's political centre, full of ornate halls and whispery chambers. Next come 10 Smithsonian museums that hold everything from the Star-Spangled Banner flag to the Apollo Lunar Module, George Washington's sword and Louis Armstrong's trumpet. The Vietnam Veterans Memorial, a heartbreaking wall of names, and the Lincoln Memorial, a temple to the beloved 16th president, anchor the west end. The Mall is also America's great public space, where citizens come to protest their government, celebrate at festivals or go for a run. The green pretty much contains the American experience in one tidy plot.

SEE IT! ***Use the Smithsonian Metro station. Most museums open 10am–5.30pm.***

© Chance Hammock / 500px

© Rattham / Getty Images

© Dove Lee / Getty Images

288

Hike the Himalaya's headiest trail in the Valley of Flowers

INDIA // High in the western Himalaya hides a 5km (3-mile) valley of almost mythological reputation. British mountaineer Frank Smythe first popularised the valley in the 1930s, calling it the most beautiful valley in the world. The real draw is botanical; a kaleidoscope of over 300 species of wildflowers carpets the meadows every July and August, cloaking the valley in a lush scent, just as the rest of India is getting pummelled by monsoon rains. Only day hikes are allowed in the national park; a second hike leads to Hem Kund, a glacial lake sacred to Sikh pilgrims.

SEE IT! *From Rishikesh in Uttarakhand state it's a day's drive to Govindghat and the trailhead at Ghangaria. Bring waterproofs.*

289

Take in Shanghai's hypermodern skyline on the Bund

CHINA // The grand sweep of the riverside Bund is symbolic of colonial Shanghai, and remains the city's standout landmark. The Bund was the city's Wall Street, a place where fortunes were won and lost. Originally a towpath used for dragging barges of rice, it has been the first port of call for visitors since passengers began disembarking here more than a century ago. These days, it's the boutiques, bars and restaurants and the hypnotising views of the Pudong skyline that pull in the crowds.

SEE IT! *The Bund Sightseeing Tunnel runs under the Huangpu River, connecting the Bund to Pudong. Best strategy? Stroll.*

© Katiekk / Shutterstock

Driving through the Empty Quarter, the desert that covers close to a third of the Arabian Peninsula

Feel insignificant amid the Empty Quarter's rippling dunes

UNITED ARAB EMIRATES // Yeah, there are other deserts on this list that have snaffled higher rankings but that's probably because more people happen to have heard about them. And you know what that means: more crowds. The Empty Quarter (Rub' al-Khali) though – well, it does what it says on the tin. This is the longest uninterrupted line of undulating sand dunes in the world, running through Saudi Arabia all the way down to Yemen. The most accessible section for visitors is the nub of the desert that nudges into the UAE.

Here are the neon-orange and rose-pink dunes of your imagination, rolling onwards to the horizon; the same unforgiving landscape that British explorer Sir Wilfred Thesiger made famous in his book *Arabian Sands*. The best way to experience this fearsome, lonely place is to spend the night. Forgo the big tourist camps with their bells-and-whistles entertainments and, instead, enlist a desert operator who'll take you out to camp in the dune field itself, where the shifting sands are the only sounds and the night sky glitters above.

SEE IT! ***Liwa Oasis in Abu Dhabi Emirate is the usual starting point for trips heading into the Empty Quarter.***

© Tomas Sereda / Getty Images

© Marco Ciannarel / Shutterstock

291

Climb the Thames-side marvel of St Paul's Cathedral

ENGLAND // St Paul's is a survivor. There's been a church here since Anglo-Saxon times, and Sir Christopher Wren's magnificent Baroque cathedral was built after the Great Fire at the end of the 17th century. Unscathed by Nazi bombs, it became a symbol of British resistance, standing tall among the smoke of the Blitz.

Just wandering the soaring interior is an uplifting experience. Steps take you up to the Whispering Gallery where, if you speak close to the wall your words will carry the 32m (105ft) to the opposite side. Staircases of stone and iron bring you to the Golden Gallery, which has wonderful views of London.

The crypt has hundreds of memorials to historic figures, including Wellington and Nelson. Wren's tomb bears a Latin inscription that simply and poignantly reads: 'if you seek his memorial, look about you'.

SEE IT! ***Admission is £17 (£7.50 for children). There are regular tours (free with admission). The nearest Tube station is St Paul's.***

292

Be dazzled by golden Kinkaku-ji

JAPAN // How do you choose an ultimate temple in a city with more than 2000 temples and shrines? Kinkaku-ji is not Kyoto's oldest temple nor the most spiritually important but it is, quite simply, one of Japan's most impressive sights. Coated in gold leaf, its reflection shimmering in the pond below, the Golden Pavilion is beautiful at any time: in autumn on a backdrop of blazing maple leaves, in winter with its curved eaves carrying a dusting of snow, or under the brilliance of the summer sunshine. Though the original structure dates from 1397, the current pavilion was restored in 1955 after a troubled and fanatical young monk burned it to the ground – a story that was fictionalised in Mishima Yukio's *The Temple of the Golden Pavilion*. Undiminished, Kinkaku-ji always attracts hordes of admirers. Try to visit early on a weekday, or later, when the temple glows at sunset.

SEE IT! ***Kinkaku-ji is in northwest Kyoto, near other prominent Unesco World Heritage temples Ryōan-ji and Ninna-ji.***

Pip Stewart's Top Five Places

Pip Stewart is a British adventurer, journalist and presenter. She has paddled the entire length of the Essequibo, South America's third largest river; cycled from Malaysia to London, and documented the deforestation of the Amazon and its terrible effect on indigenous communities.

THE JURASSIC COAST, UK – It's amazing how much world-class scenery and adventure is on our doorstep. The hiking paths on the Jurassic coast and Pembrokeshire coast are both definitely worth getting sweaty for.

THE SCOTTISH HIGHLANDS – For those into dramatic landscapes, Scotland is stunning, especially the Highlands and the surrounding islands. Hiking into the wilderness is the best way to feel the raw beauty of the landscape.

AMAZON JUNGLE – If you want to feel fully alive, spend time in the jungle. Its cacophony of sounds and smells make you feel very small in the grand scheme of nature. Make sure you visit with local, ethical tourist providers.

TBILISI, GEORGIA – Artsy Tbilisi completely stole my heart. Georgia's vibrant capital is a visual delight, from quirky sculptures to art nouveau buildings next to statement modernism. Don't miss Georgian wine and the cheese-loaded bread called *khachapuri*.

UZBEKISTAN'S CYCLING TRAILS – I can't recommend human-powered journeys enough. Whether you walk, cycle or paddle, going somewhere under your own steam leaves you feeling on top of the world (and I guarantee you'll have better stories, too).

Soak up some of India's most sublime Buddhist cave art in Ajanta

INDIA // It takes time for your eyes to adjust to the light inside the 2000-year-old caves of Ajanta but, slowly and surely, the stunning detail of the carved stone and painted murals starts to appear: of serene Buddhas and princely bodhisattvas, and detailed scenes depicting stories from the Buddha's past lives. Like many of India's great Buddhist sites, the caves pockmarking the horseshoe-shaped gorge lay deserted for over a millennium, until a British officer stumbled across them quite by chance in 1819. Today the collection of caves, ancient monasteries and prayer halls ranks as one of India's greatest treasure houses of Buddhist art.

SEE IT! ***The caves are 105km (65 miles) from the Mughal city of Aurangabad in Maharashtra state, and easily visited from Mumbai (from where you can also visit the world-class cave temples of Ellora).***

© DZarzycka / Getty Images

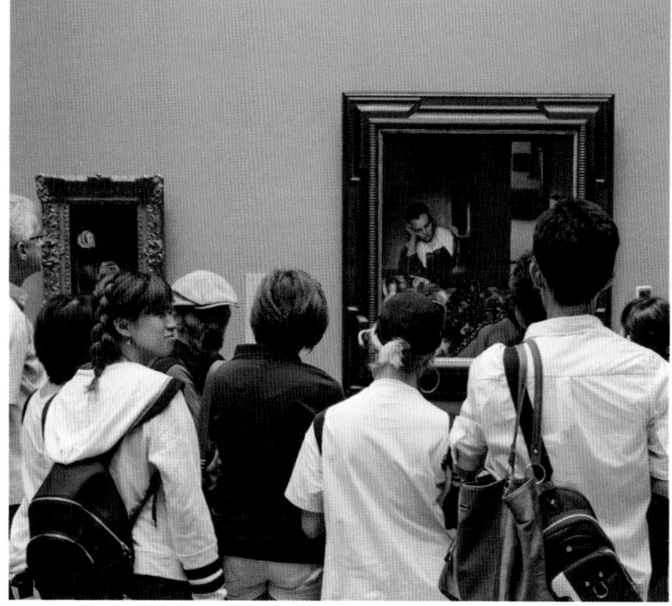
© Guillaume Gaudet / Lonely Planet

294

Walk with pilgrims at the Church of the Holy Sepulchre

ISRAEL & THE PALESTINIAN TERRITORIES // Venerated as the site of Jesus of Nazareth's crucifixion, as well as his tomb, Jerusalem's Church of the Holy Sepulchre is one of the holiest places in the world for Christians. The Via Dolorosa, a walking path threading together sites associated with Jesus' death and resurrection, culminates inside the church, which has been a focal point for pilgrims from around the world for more than 16 centuries. The keys to the church remain under guardianship of Jerusalem's oldest Muslim family, preventing different Christian factions from vying for control of this cherished sanctuary. It is an awe-inspiring complex, even in a city that's full of places of worship, attracting the crowds and the intensity to match.

SEE IT! *The church is in the Christian Quarter of Old Jerusalem. Dress modestly, and expect a long queue for the tomb.*

295

Ogle masterpieces at the Metropolitan Museum of Art

USA // The Met's ability to thrill, confound, inspire and exhaust with its coffer of art and antiquities has made it one of the world's most popular museums. More than six million visitors a year come to gawk at its unrivalled collection of ancient Egyptian art, European and American paintings, Greek sculptures, African and Oceanic masks and medieval arms – and that's just a snapshot of the treasures within. It's a self-contained cultural city-state with the world in its palm and when we say it's big, we mean it: at nearly 7 hectares (17 acres), it's H-U-G-E. Don't miss the Temple of Dendur, complete with its 2000-year-old stone walls covered in hieroglyphics. Museum tickets are good for three consecutive days, but you'll still only see a smidge of the trove.

SEE IT! *The Met is on New York City's Upper East Side, reachable by subway or bus. Avoid weekends if you dislike crowds.*

© Maciej Bledowski / 500px

Perform a walking meditation at Bodhnath's great stupa

NEPAL // It would not be an exaggeration to say that life in Kathmandu revolves around Bodhnath. For thousands of Nepali Buddhists, every day starts and ends with a ritual circumambulation of the country's largest Buddhist stupa. Joining this human tide, and spinning the lines of prayer wheels while gazing on Bodhnath's golden Buddha-eyed tower, is an electrifying and moving experience – an interactive act of pilgrimage and people-watching in which you are both an observer and participant. Bodhnath is the heart of Nepal's international Buddhist community, and the streets around the stupa are lined with shops offering yak butter, prayer flags and relics. Nowhere else in the country exudes such a sense of devotion.

SEE IT! *Bodhnath is 7km (4.3 miles) from central Kathmandu and ringed by guesthouses and restaurants; come on full moon nights to see the stupa illuminated by thousands of butter lamps.*

© Blaine Harrington III / Getty Images

Feel the cruelty of the recent past at the Apartheid Museum

SOUTH AFRICA // Johannesburg's Apartheid Museum mixes historical detail with a visceral treatment of apartheid, the repressive system of racial segregation that tore 20th-century South Africa apart. From the start, visitors empathise with people confronted with institutionalised racism in every aspect of their lives. You are given an arbitrary 'racial classification' card before entering the building through gates marked 'whites' and 'non-whites' – recalling the signs that covered South African public spaces from park benches to beaches.

The museum profiles both leaders, such as HF Verwoerd, 'architect of apartheid', and everyday people who struggled for dignity. Apartheid only ended in 1990, when Nelson Mandela was released from prison.

SEE IT! *The Apartheid Museum is located 8km (5 miles) south of Johannesburg city centre, just off the M1 freeway. Get there by Uber or rental car.*

© Justin Foulkes / Lonely Planet

© Matt Munro / Lonely Planet

298

Face the world's most extreme tides at the Bay of Fundy

CANADA // The highest tides on the planet sweep this inlet between New Brunswick and Nova Scotia, giving the place a schizophrenic personality. Boats look hopelessly beached on mud flats, and piers look ridiculously lofty at low tide. Then, when over 100 billion tonnes of water flow back into the bay, boats bob up to 15m (49ft) above where they were six hours earlier – it's surreal. (If you're having trouble conceiving of the water amount, think of it this way: enough enters in one tide cycle to fill the Grand Canyon.) Get into the action by rafting the tidal bore or taking a whale-watching cruise. The latter is exceptional here, as the tides stir up serious krill and plankton that draw blue, minke and rare fin whales.

SEE IT! *St Andrews, New Brunswick and Advocate Harbour, Nova Scotia make good bases. Check www.tides.gc.ca for tide times. Fundy's whale watching season is from June to October.*

299

Get salty underground at Wieliczka Salt Mine

POLAND // Descend deep into a strange, dimly lit netherworld, full of artistic, architectural and spiritual treasures – all made of salt. Wieliczka operated as a salt mine from the 13th century all the way to 1996 and has been welcoming tourists for nearly 300 years. Reaching depths of 327m (1072ft), the mine has 287km (178 miles) of tunnels. Its labyrinthine reaches contain chapels, a saltworks museum, hundreds of salt sculptures and much more. The highlight is the gargantuan Chapel of St Kinga, with its ornate salt chandeliers and altarpiece. There's also an underground lake and a 'subterraneotherapy' spa, which takes advantage of the purported health benefits of the mine's unique microclimate. All we can say is, it's surreally salty down there.

SEE IT! *Wieliczka is located 14km (8 miles) from Kraków, Poland. No matter the weather outside, it's chilly when you're 135m (443ft) underground.*

300–399

300

Ace Egyptology inside the Grand Egyptian Museum

EGYPT // Yes, it is cheeky including a museum which hasn't even opened yet. But when it does finally unlock its doors – fingers crossed for 2020 – Giza's Grand Egyptian Museum (GEM) promises to become one of the world's great institutions. The largest museum ever devoted to one civilisation, this is where you'll come to understand the scope of Egypt's pharaonic history and gawk in mind-boggling wonder at the full caboodle of treasures unearthed in Tutankhamun's tomb.

SEE IT! ***The GEM is taking shape in Giza, 9km (6 miles) from downtown Cairo and not far from the Pyramids of Giza main entrance.***

301

Scale Mordor on Piton de la Fournaise

RÉUNION, FRANCE // The French island of Réunion owes a lot to the volcano known as Piton de La Fournaise ('peak of the furnace'). For one thing, a significant proportion of the isle spewed from its summit as molten magma. Despite the paradisiacal setting, this is one of the world's most active volcanoes, with more than 100 catalogued eruptions since 1640. If you've seen the depiction of Mordor in Peter Jackson's *Lord of the Rings*, you'll have some idea what to expect on a trek here.

SEE IT! ***The village of Bourg-Murat is the gateway to Piton de la Fournaise; trek across the tortured lava to the Dolomieu, the latest and most active cone.***

© Milosz Maslanka / Shutterstock

303

Feel Vienna's pulse at the avant-garde MuseumsQuartier

AUSTRIA // If you needed proof that Vienna is at art's cutting edge, the MuseumsQuartier is it. This remarkable hub brings together galleries, museums, restaurants and bars in the vast former imperial stables. Showstoppers include the world's largest collection of Egon Schiele paintings in the Leopold Museum, and modern art powerhouse MUMOK, its collection as serious as its brooding basalt facade. It also harbours top kids' museum Zoom and contemporary dance venue Tanzquartier.

SEE IT! ***MuseumsQuartier is best tackled by purchasing a combined ticket that includes entry to every museum.***

302

Lift your spirits on Scotland's Northeast 250

SCOTLAND // The majesty of Scotland's west coast is undeniable, but it's never been the whole story. The Northeast 250 driving and cycling itinerary connects Scotland's most famous whisky region, rich farmland, castles, sandy beaches and mysterious Pictish stones. Its highlight is probably the Cairngorms, Britain's greatest mountain range, where pretty villages sit below towering giants of granite and heather. The route shines a light on a corner of Scotland where idyllic glens and astounding sea views combine – sights to lift anyone's spirits.

SEE IT! ***The Northeast 250 can be tackled by car or bike, as quickly or slowly as you like – there's plenty of accommodation en route.***

304

Soak up the cinematic landscape of Ao Phang-Nga

THAILAND // Imagine turquoise bays out of which shoot limestone rock towers draped in lush tropical greenery, and dazzlingly white beaches bracketed by picturesque fishing villages. Phang-Nga Bay is perhaps Thailand's most exquisite landscape, and one that is instantly familiar, thanks to its starring role in the James Bond film *The Man with the Golden Gun*. Your mission is simple: to kayak, rock climb, sail around on a yacht or live out your own island fantasy swinging in a hammock.

SEE IT! ***From November to April the bay is busy. Visit early morning (ideally from the Ko Yao islands) or stay later to avoid the crowds.***

305

Be moved by the culture of Titicaca's Reed Islands

PERU // Meeting the Uros people of Lake Titicaca is a moving encounter – partly because the artificial floating islands they live on bob a bit, but mostly because theirs is an existence that's absolutely unique. The Uros pre-date the Incas, and have inhabited this group of islands made from dried totora reeds for centuries. As the reeds at the bottom rot, new ones are added on top; the Uros also use the reeds to construct their boats and houses. The islands feature a reed watchtower and traditionally, if a threat was spotted, the islands could be moved.

SEE IT! *The gateway town is Puno, Peru, accessible from Lima by bus and from Cuzco and Arequipa by bus and train.*

© saiko3p / Shutterstock

306

Wander back to Roman times at Diocletian's Palace

CROATIA // Many cities have Roman ruins, but in Split they form the fabric of the city. The palace, built by the emperor Diocletian, is at the heart of this lively port. Diocletian imported marble from Italy and Greece, and sphinxes from Egypt for his grand retirement home. It's a wonderful place to wander. The palace has more than 200 buildings and houses some 3000 people. Some passageways are deserted and others thump with music from bars and cafes, while the residents hang out their washing and kids play football between the ancient walls.

SEE IT! *There's no charge to enter the palace. Split is served by direct flights from many European cities, and by ferries along the coast.*

© Kirk Fisher / Shutterstock

307

The artfully lit baroque terraces around Salamanca's Plaza Mayor dazzle after dark

Ponder local life in Salamanca's Plaza Mayor

SPAIN // There are innumerable Plaza Mayors in Spain, from the terracotta terraces of Valladolid to the rectangular facades of Madrid, but none are as extraordinary and spectacular as the broad central square in Salamanca. Rendered in golden sandstone and designed in the second quarter of the 18th century by Catalan architect Alberto Churriguera, this giant plaza is ringed by an unbroken ensemble of late-baroque terraces held up by 88 arches and adorned with 247 balconies. The overall effect is at once harmonious and monumental. Bullfights were held here from the plaza's foundation until well into the 19th century. These days the spectacles are mercifully less gory: classical concerts, meetings of old amigos, cafe crawls, bar blitzes and city tours. Slump down in one of the many alfresco cafes that ring the perimeter and admire Señor Churriguera's masterful ornamentation over a café con leche. The square has many moods. At sunset, slanting rays give the ubiquitous sandstone a warm romantic hue. At night, clever illumination lends the balconied mansions a glowing ethereal quality.

SEE IT! ***The square is 10 to 15 minutes' walk from Salamanca's train and bus stations.***

308

Clamber through Postojna Cave and Predjama Castle

SLOVENIA // Slovenia's subterraneous Postojna region has an enviable swag bag of attractions. First, there's brooding Predjama Castle, wedged into the gaping mouth of a cavern halfway up a 123m-high (404ft) cliff, looking unconquerable. Drawbridge? Tick. Dank dungeon? Tick. Ancient treasure haul? Tick. Less than 10km (6 miles) down the road, the awesome karst cave of Postojna swallows crowds in a series of caverns, halls and passages some 20km (12 miles) long and two million years old. These subterranean systems harbour the unusual and totally blind olm – a vertebrate endemic to southeastern Europe's karst, and the world's largest exclusively cave-dwelling animal.

SEE IT! *Postojna is an hour southwest of Ljubljana by bus.*

© Sushil Kumudini Chikane / Shutterstock

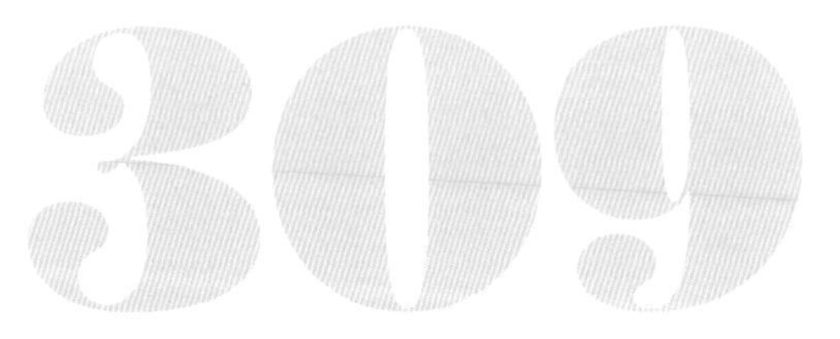

309

Discover Laos' answer to Angkor at Wat Phu

LAOS // Laos is better known for laid-back escapes than ancient ruins, but at Wat Phu, you get both. This formidable complex of Khmer-era temples is tucked away in pristine jungle on the bank of the Mekong River. The tumbledown pavilions, graceful carvings and tall trees give Wat Phu a mystical atmosphere, and it's still a place of worship today. While it lacks the enormity of Angkor Wat, the few visitors and dramatic natural setting help make Wat Phu an impressive, soulful alternative.

SEE IT! *Pakse, the capital of Champasak province, is the closest town. Visit early for cooler temperatures and the best light.*

310

Track Bengal tigers in Tadoba-Andhari Tiger Reserve

INDIA // A twig snaps deep in the forest. Everyone freezes and your pulse starts to race, as you try not to breathe or wipe the sweat from your eyes. Probably it's a bison-like gaur. But maybe, just maybe, you'll hear a roar and glimpse a blur of tiger stripes in a moment of such primeval intensity that it will leave you shaking. And with the numbers of India's famed big cats now double that of 2011, your odds of a sighting have never been better.

SEE IT! *Tours explore the park in 4WDs, on foot or in canoes, always in the company of local guides.*

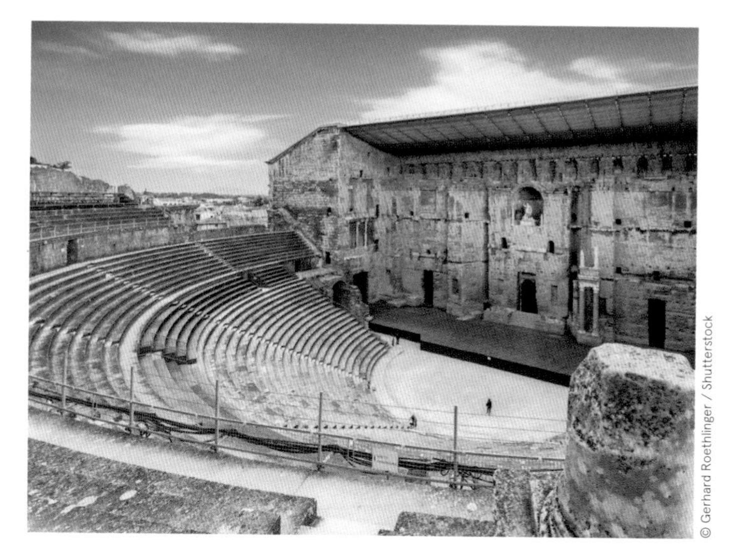

© Gerhard Roethlinger / Shutterstock

© Jad Davenport / National Geographic Image Collection / Alamy Stock Photo

311

Behold the Romans' Théâtre Antique in the Provençal town of Orange

FRANCE // In Provence's Vaucluse region, famed for its lavender and vineyards, the town of Orange was once one of the Gallo-Roman empire's major settlements. Legacies from its heyday include a mighty 1st-century-AD triumphal arch commemorating Roman victories in 49 BC. But the star of the Roman show is the ancient theatre, the Théâtre Antique, once the largest in Gaul and now the best preserved in Europe, which saw it listed as a Unesco World Heritage site.

Built from limestone, its proportions are typically monumental, with a stage wall stretching 37m (121ft) high, 103m (338ft) wide and 1.8m (6ft) thick (Louis XIV extolled it as the finest wall in his kingdom), and a capacity of 10,000. Not simply a relic, it hosts performances that still enthral audiences in its spectacular surrounds.

SEE IT! *Trains link Orange to transport hubs including Avignon, Arles, Vitrolles station (for Marseille-Provence airport) and Marseille.*

312

Make a sacred spelunk into Actun Tunichil Muknal

BELIZE // A 45-minute hike through lush jungle brings you to the opening of Achtun Tunichil Muknal, the Cave of the Stone Sepulchre (through which claustrophobics ought not pass). The Maya believed it was a portal into the underworld, a chasm filled with rivers of blood and scorpions. Summoning courage and donning a headlamp, you swim across an icy pool guarding the entrance. Through inky blackness you walk, climb and crawl, passing shimmering rock formations and dripping stalactites, until you reach the caverns where thousand-year-old skeletons met their doom in ritual Mayan sacrifices. Pay your respects and continue to the main chamber. Inside lies the calcite-encrusted remains of the Crystal Maiden. She's waited for millennia, so be reverent. This is among Belize's – and the Maya's – most sacred spaces.

SEE IT! *The round trip from San Ignacio (including the drive into the jungle) takes 10 hours and can only be done with a licensed guide.*

313

Promenade through Timgad's lonely Roman ruins

ALGERIA // A sprawl of bathhouses, toppled temples and chapels, separated by crisscrossing colonnaded streets, Timgad is a vast, ruined Roman settlement extraordinaire. Sure, this list has plenty of Roman sites – and popular ruins like Pompeii and the Colosseum have poached higher positions – but Timgad has its own trick up its sleeve. There's no one here. This is one of the best-preserved Roman cities in the world and it's free from crowds.

Founded in approximately AD 100 by the Emperor Trajan, Timgad was conceived as a military colony to boost Roman presence and power in the region and subdue the local Berber population. Originally designed as a perfect square (it remains a superb example of the Roman Empire's obsession with urban planning), it later lurched outside the boundaries of its symmetrical outline. From military settlement beginnings, Timgad struck its heyday in the 2nd and 3rd centuries, becoming a prosperous Christian centre and bishopric and adding large elegant residences and churches to its mix. The good times unfortunately didn't last. The Vandals attacked in AD 430 and after a 5th-century raid by local Berber tribes the city fell into decline.

Standing amid the ruins, you can fully imagine Roman life here. There's so much to explore but make time to walk down the Cardo Maximus to the Forum and sit for a while in the Theatre to survey the sweep of the perfect grid laid out below. Then veer west along the Decumanus Maximus to stare up at Trajan's Arch, which soars over the half-collapsed walls. Just remember: shhhh. Don't tell anyone about it.

SEE IT! ***Timgad is in Algeria's Aurès Mountains region. Most people visit from Constantine, 120km (75 miles) away. Don't miss Timgad's museum while there.***

© Anton_Ivanov / Shutterstock

© eXpose / Shutterstock

The opulent Great Hall at Christ Church, where teen wizards sat down to dine in scenes from the Harry Potter films

314

See old masters and young wizards at Christ Church

ENGLAND // Albert Einstein, Lewis Carroll, Joshua Reynolds, John Locke and 13 British prime ministers all studied at Christ Church. A few of these famous names might be turning in their graves to know that this august institution is best known today as the setting for several scenes in the Harry Potter films.

But if the academic gods didn't want this Oxford college to provide the backdrop for a fantasy saga perhaps they shouldn't have made it so picture perfect. Its honey-coloured stone is wonderfully pleasing on the eye, and neatly manicured Tom Quad, overlooked by imposing Tom Tower, is so large that Royalist soldiers used it as a cattle pen during the English Civil War. The Renaissance Great Hall (the inspiration for the Great Hall of Hogwarts) is simply magnificent, while its cathedral mixes Norman and Gothic features beneath a harmonious vaulted ceiling.

Elsewhere, the Picture Gallery has works by Tintoretto, Michelangelo, da Vinci and Raphael, while Christ Church Meadow is a lovely green space, bordered by the Cherwell and the Thames and inhabited by a herd of longhorn cattle.

SEE IT! *Trains from London take just over an hour. The college is generally open to visitors 10am to 5pm (from 2pm on Sunday).*

315

See divine forests and dolphins at Golfo Dulce

COSTA RICA // With a name aptly translating as 'Sweet Gulf', the Golfo Dulce is southern Costa Rica's vast, biodiverse inlet formed by the hammerhead-shaped Península de Osa (which encompasses Parque Nacional Corcovado, also on this list at #74). The divine tropical forest and lowland wet forest tract stretching around the inlet, rimmed by Pacific Ocean surf, includes the Golfo Dulce Forest Reserve and Parque Nacional Piedras Blancas, which are home to some wonderful wilderness lodges. Recent efforts to protect waters around the gulf have won international acclaim for providing the scalloped hammerhead shark, humpback whale and four dolphin species with a safe haven.

SEE IT! *Most lodges will pick up from Puerto Jiménez (west side) or Golfito (east side) airports, which connect to San José.*

Sarah Marquis' Top Five Places

Sarah Marquis is a Swiss adventurer and explorer. From 2010 to 2013, she walked 20,000 kilometres alone from Siberia to the Gobi Desert, into China, Laos, Thailand, and then across Australia.

01

THE THREE CAMEL LODGE, DALANZADGAT, GOBI DESERT – I am completely smitten by this place and remember following the wise Bactrian camels in my search for water then waking up with a wolf howling around my tent.

02

THE BIBBULMUN TRACK, WESTERN AUSTRALIA – This trail runs for 1000km and I've walked it more than five times, usually as training prior to an expedition. It is a truly special place.

03

LAKE BAIKAL, SOUTH SIBERIA – The lake freezes in winter and everything feels very other-worldly. The waters are filled with nerpa (an elusive type of freshwater seal found only in Siberia).

04

'THE FINISHING TREE' – This is my tree in south Australia, west of the Nullarbor plain. It marks the point where I finished my big expedition. Some people even go and pay tribute to it now.

05

GREAT SANDY DESERT, WESTERN AUSTRALIA – This is the most beautiful desert that I've crossed. At almost 3000 sq km, there is nothing around but thousands of dunes and intense red sand. It's a special place for many aboriginal people, too.

© Matthew T. Carroll / Getty Images

316

Reflect on loss and hope at the National September 11 Memorial & Museum

USA // Rising from the ashes of Ground Zero, where the World Trade Center's twin towers crashed down, this New York City memorial is a dignified response to the deadliest terrorist attack on American soil. Occupying the actual footprint of the ill-fated buildings, two reflecting pools now weep like dark, elegant waterfalls. Framing them are the names of nearly 3000 people who lost their lives on 11 September 2001 and in the 1993 World Trade Center bombing. The Memorial Museum lies beneath the plaza and digs deeper into the tragic events. You'll flip-flop emotionally between grief (hearing voicemail messages left by victims to loved ones) and hope (hearing survivor stories) among the exhibits.

SEE IT! ***The memorial and museum are in Lower Manhattan, easily accessible by subway trains. The memorial is free; the museum charges admission but is free Tuesday evenings.***

© Christian Wittmann / Shutterstock

317

Search for the ghosts of the ancient world at Butrint

ALBANIA // Ancient Greeks and Romans both left their mark on Butrint, a fortified trading city and subsequent Byzantine ecclesiastical centre that now lies in hugely atmospheric ruins, secluded in a verdant forest at the heart of a national park on Albania's Ionian coast. The remains of an acropolis, a 3rd-century-BC Greek theatre, a grand Roman villa, and public baths decorated with geometric mosaics are just some of the crumbling gems you'll encounter here. The relief of a lion killing a bull on one of the ancient stone gates is redolent of the long-gone assailants from Butrint's history. Tranquil Lake Butrint makes a perfect backdrop to the archaeological site.

SEE IT! ***The ruins lie 18km (15 miles) south of Saranda. The mosaics are covered with sand for protection from the elements and can usually be seen only between mid-August and mid-September.***

318

Trek to the world's highest peak at Everest Base Camp

NEPAL // Nepal's most famous trek takes you through Sherpa villages and colourful monasteries into the heart of the grandest mountain range on earth. It's the perfect combination of close-up mountain splendour and cappuccino-infused trekking-lodge comfort. Joining the trekkers in April and May are a stream of mountaineers and Sherpa guides, which only adds to the colour. Avoid the crowds and it's hard to think of a more spectacular region for trekking.

Everest has at times become a victim of its own success, with overcrowding on the mountain creating deadly bottlenecks, and litter becoming a major problem. You can do your bit by purifying your own water and supporting the recent ban of single-use plastics in the region.

SEE IT! *Fly from Kathmandu to Lukla and start trekking from the runway. Everest views are best October to mid-December.*

© Meiqianbao / Shutterstock

319

Follow in the footsteps of saints at Studenica Monastery

SERBIA // In a country bursting with Byzantine-style monasteries, Studenica stands out, and for good reason: it was established in 1196 by the founder of the Serbian empire (and future saint), Stefan Nemanja. Within the walls of the ancient complex, hushed monastic life has thrived throughout the turbulent centuries. The grand white-marble Church of Our Lady is the final resting place of Stefan Nemanja, while the elegant King's Church is a showcase of ethereal medieval frescoes. To truly walk in the saints' steps, visit the hermitage of St Sava (Nemanja's son and revered founder of the Serbian Orthodox Church), a cave-like clutch of monks' cells carved into a cliff soaring above the Studenica canyon.

SEE IT! *Studenica is 40km (25 miles) southwest of Kraljevo. St Sava, accessed via a footbridge, is 12km (7 miles) from Studenica.*

© Artem Mishukov / Shutterstock

© BTWimages / Shutterstock

© ESB Professional / Shutterstock

320

Celebrate Roman craftwork at the Bardo Museum

TUNISIA // Home to one of the world's most magnificent, must-see collections of Roman mosaics, the Bardo is Tunisia's most important museum. Inside, a vibrant vision of ancient North African life is presented in glorious detail, thanks to the superbly well-preserved haul of mosaic art. Few original mosaics remain in situ in Tunisia's archaeological sites, having been excavated and brought here for preservation, making a visit to the Bardo a necessity before exploring the country's impressive collection of Roman ruins.

SEE IT! *The Bardo is 4km (2.5 miles) northwest of central Tunis and can be reached by tram. Take tram Line 4 and get off at the Bardo stop.*

321

Ride the rickety tram to Hong Kong's Peak

CHINA // Emerging from a tramcar at the Peak is like leaving a sauna, such is the relief from the blanket of humidity that smothers Hong Kong city. The funicular railway has ferried people up here for 130 years, to escape the heat and gulp in panoramic views. Watch black kites calmly gliding on thermals, while the metropolis below goes about its frenetic business around Victoria Harbour. There are observation decks at Sky Terrace 428 and Peak Galleria. To fully absorb your surroundings, do the 3.5km (2-mile) Peak Circle Walk.

SEE IT! ***Riding the Peak Tram on its vertiginous route past HK's mega modern skyscrapers is an experience in its own right.***

With its twin towers and rooftop dish and dome, the Congresso Nacional is one of Brasília's best-loved buildings

322

Explore the architecture of utopia in Brasília

BRAZIL // Urbanists still use Brasília as an example when discussing the possibilities and pitfalls of purpose-built cities. Because unlike Rio or São Paulo, Brasília did not develop slowly over time. It was designed and erected in just a few years in the late 1950s to serve as Brazil's new capital. The result is an odd but fascinating metropolis with none of the colourful hustle-bustle of its coastal cousins. Massive modernist buildings like something out of *The Jetsons* are planted along imposingly wide roads more fit for jet cars than pedestrians. Grab your camera for an architectural tour of such structures as the Cathedral of Brasília, designed by starchitect Oscar Niemeyer and resembling a crown for a space emperor. Gawk at the superquadras: huge residential blocks with shared shopping, education and recreational facilities. Dated-looking now, they were envisioned as a new kind of utopian living space. Though many Brazilians dismiss the city as dull, it's got a lauded food scene, excellent museums and no end of peaceful green spaces.

SEE IT! ***Brasília's connected to most Brazilian cities by air and bus, though getting to Rio or São Paulo overland will take most of a day.***

323

See elephants and snow in Amboseli National Park

KENYA // There's no better place on Earth to watch elephants than Amboseli National Park in Kenya's south. Africa's highest mountain, the snow-capped Mt Kilimanjaro, acts as backdrop to the magical scenes at dawn or dusk when the clouds part and the elephants come out to play. Better still, Amboseli was spared the worst of Kenya's past poaching crisis and the big-tusked elephants are remarkably tolerant of humans, allowing you to get closer than elsewhere.

☛ SEE IT! ***Rent a vehicle (with driver) from Nairobi or Mombasa. Avoid the March to May rainy season, when wildlife disperses.***

Take a holy trip to Santiago de Compostela cathedral

SPAIN // With a decorative Churrigueresque facade beckoning brightly to exhausted hikers hallucinating after weeks on the Camino de Santiago, Santiago de Compostela's glorious cathedral is a sight for sore eyes – and legs! Herein, legend decrees, lie the remains of St James the apostle. When you've finished admiring the famous Obradoiro facade (so famous it's imprinted on Spain's euro coins), duck inside to lap up the contrasting Romanesque interior, the largest of its kind in Spain and the terminus of Europe's most popular pilgrimage.

☛ SEE IT! ***Santiago de Compostela is well connected by bus, train and plane – plus the multi-day Camino de Santiago trek.***

325

Ponder the scars of the past from Stari Most

BOSNIA & HERCEGOVINA // Mostar's magnificent Ottoman-era bridge spans the turquoise Neretva River, its fortunes mirroring those of Bosnia. The town was the front line during the heart-rending 1990s civil war, and the bridge was destroyed. It was painstakingly reconstructed in 2004 and, like few other structures on Earth, it serves as a powerful icon of reconciliation. Stay overnight (or visit between October and April) to admire it without the summer day-tripping crowds.

☛ SEE IT! ***Watch the Mostar Diving Club continue a centuries-long tradition – jumping off the bridge into the Neretva, 24m (79ft) below.***

Browse Europe's treasures at the Rijksmuseum

NETHERLANDS // The Rijksmuseum stuffs masterpieces galore into its 1.5km (1 mile) of galleries. The crowds huddle around Rembrandt's humongous *Night Watch* and Vermeer's *Kitchen Maid*, and who can blame them? Both paintings are Golden Age icons. But step into the 100 or so rooms beyond, and the most impressive treasures await: antique ship models, savage-looking swords, crystal goblets and magic lanterns from the 17th century. You could spend days gaping at the beautiful and curious collections tucked into the nooks and crannies.

☛ SEE IT! ***The Rijksmuseum is in Amsterdam's Old South district. Queues can be lengthy; book online for fast-track entry.***

Pōhutu Geyser, said to have been the work of Māori fire goddesses Te Hoata and Te Pupu

327

Get steamed up in the geothermal wonderland of Whakarewarewa

NEW ZEALAND // Nicknamed 'Sulphur City', Rotorua is a land shrouded in steam, thanks to geothermal activity beneath the central North Island's thin crust. It's long been a centre for the Māori, who make up more than a third of the population. In Māori legend, the Goddesses of Fire, Te Hoata and Te Pupu, emerged from the earth here, leaving volcanoes, geysers and hot springs in their wake.

South of downtown Rotorua, Whakarewarewa is a living Māori village, where *tangata whenua* (local people) have resided for centuries, and Mother Nature displays her fiery side in a frenzy of steaming vents, boiling mud pools and unpredictable geysers. Visitors feel the heat on walkways, guided by villagers who tell tales of old and demonstrate traditional Māori arts, including the famous haka, singing, flax weaving and carving. Nearby, the Tarawera Trail winds for 15km (9 miles) through native bush to Hot Water Beach, where you can camp on the shores of Lake Tarawera and soak your trekking-sore feet in a natural hot spring.

SEE IT! ***Rotorua is three hours' drive south of Auckland on the Thermal Explorer Highway, a themed road trip taking in many other major North Island highlights.***

328

Dive among turtles and rays in Malaysia's underwater Eden

MALAYSIA // Located 36km (22 miles) off Sabah's southeast coast, Sipadan sits atop an extinct volcanic cone surrounded by near-vertical walls. Blanketed in technicolour corals, they form a veritable way station for all types of sea life, from barracuda to bumphead parrotfish and, from around March to May, whale sharks. Sea turtles and reef sharks are a given during any dive, and divers might also spot mantas, eagle rays, octopus and scalloped hammerheads. Book ahead – only 120 Sipadan passes are issued each day.

SEE IT! *Nearby Mabul island is the region's accommodation centre; you can't sleep over on Sipadan.*

View visionary works at Louisiana Museum of Modern Art

DENMARK // It's not only what's inside this long, low, white building – a masterpiece of modernist Danish design from the 1950s – that guarantees its place on this list. Louisiana's location, right on the shore of the Øresund coast, makes it unmissable. When you've meandered through one of Scandinavia's most important collections of modern art (Picassos and Hockneys included), head out to explore a sculpture park packed with works from the likes of Henry Moore, or simply dangle your toes in the sea.

SEE IT! *From Copenhagen, take a 35-minute train ride to Humlebæk. The museum is a short walk from the station.*

330

Tour the apocalypse in wild, haunting Chernobyl

UKRAINE // Decades after Chernobyl's reactor core went into meltdown in 1986, the world's worst nuclear accident has become an unlikely tourist attraction. Tours of the exclusion zone offer up-close views of the apocalypse – entire towns abandoned in the midst of the catastrophe. It's incredibly eerie and sobering, as wild deer, wolves and elk wander through the wasteland. Forget about visiting the reactor – it's still a radioactive hotspot and no one is allowed within 200m (656ft) of the site.

SEE IT! *Tours from Kyiv run to Chernobyl inside the exclusion zone; radioactivity is carefully monitored throughout the trip.*

328

331

Seek out manatees, mangroves and migratory birds in the otherworldly landscape of the Lençóis Maranhenses.

Walk the dunes and lagoons of Parque Nacional dos Lençóis Maranhenses

BRAZIL // Surely this can't be Planet Earth? A first glimpse of the Lençóis Maranhenses is like landing on another world. Of all Brazil's natural spectacles, this 70km-long (40-mile), 25km-wide (15-mile) expanse of dunes resembling *lençóis* (bed sheets) spread across the landscape has to be the most unexpected.

From around March to September the dunes become pockmarked by thousands of crystal-clear, blue lagoons. They're created when rainwater fills the dunes' hollow valleys and is trapped by the impermeable rock beneath the sand. The result is a unique terrain that can be visited by 4WD, boat or on foot. At the park's Atlantic edge, stretch out on blindingly white beaches and wade through mangrove forests, looking for manatees and migratory birds like scarlet ibises.

SEE IT! ***Lençóis Maranhenses is in Brazil's northeastern state of Maranhão. The lagoons are at their best in July and August.***

Marvel at the solid rock Drakensberg Amphitheatre

SOUTH AFRICA // There's some poetic licence in the Amphitheatre's name, as befits its location in the vast range known as the Drakensberg (Dragon Mountains) in Afrikaans and uKhahlamba (Battlement of Spears) in Zulu. This is no Roman structure, but a natural rock arena where the Tugela Falls drop 850m (2788ft) in five stages (in winter, the top one can freeze over). It's a sublime 8km (5 miles) wall of cliff and canyon that's spectacular from below and even more so from on high. A popular route to the top is the Sentinel Peak day hike, but note that it involves scaling sheer cliffs on chain ladders. The Amphitheatre is the crowning glory of the Royal Natal National Park, where rustic camps, great hiking and horse trails abound.

SEE IT! *The park abuts the northern border of Lesotho, about four hours' drive from Durban. The car park for the Sentinel Peak hike is accessed south of Phuthaditjhaba, Free State.*

Tony Giles' Top Five Places

Tony Giles is a blind British solo traveller and author of e-books Seeing the World my Way *and* Seeing the Americas my Way .

01

ANTARCTICA – Being there with laughing penguins and grunting elephant seals is magical and the stench of guano and decaying seaweed takes your breath away. It exposes all the senses, with silence that rings in the ears.

02

WINCHESTER, UK – This well constructed city has a historic Norman cathedral, the Winchester Great Hall and King Arthur's Round Table (now believed to be a fake). Don't miss the delightful sound of the River Itchen as it tinkles through channels in the city centre.

03

PAPUA NEW GUINEA – Wild and rugged. Visit the north coast of the mainland and the highlands, parts of which remain largely unexplored. People are kind, have a fascinating culture and are extremely Christian.

04

DUBLIN – Temple Bar might be too touristy for some, but it's lively! Dublin has so many landmarks and statues, some of which can be touched. Dublin Castle with its long and colourful history is also very accessible.

05

IGUAZÚ FALLS – A must-visit straddling the Argentine-Brazil border. Imagine standing on a gantry above water crashing down the Iguazu River. The sound is like thunder. Or take a speedboat for a drenching in the world's widest falls. It's a blind traveller's dream.

© Christian Heeb / Prisma by Dukas Presseagentur GmbH / Alamy Stock Photo

© Antoine Barthelemy / Shutterstock

333

Lose track of time on the Copper Canyon railway

MEXICO // Mexico is a country with pitifully few passenger trains, but the one loco it does have plies what is perhaps the most spectacular stretch of railroad on earth. Incorporating 37 bridges, 86 tunnels and a conveyor belt of nose-to-the-window scenery that switches from cactus-studded desert to flowery meadows to graceful pine trees, the Copper Canyon railway is a mind-boggling feat of engineering and reason alone to brave the tetchy atmosphere of Northern Mexico. Four times larger than the Grand Canyon and deeper to boot, the rough and ready melange of gorges and plateaus is the dusty domain of the legendary Tarahumara. The indigenous people set up craft stalls beside the track but traditionally eschew train travel for two-legged transport: they're famed for their long-distance running feats.

SEE IT! ***The port of Los Mochis is the departure point for the train, book a month ahead for peak-season travel.***

334

Mural-spot in Medellín's La Comuna 13

COLOMBIA // Long one of the highest-crime neighbourhoods in Medellín, La Comuna 13 is now considered an urban success story. Tenaciously clinging to the side of the city's western mountains, it's gone from no-go zone to creative hub as drug crime decreased and infrastructure investment increased. Its famous outdoor escalators, built to help residents up the steep streets, are lined with startling street art – goofy cartoon animals, intricate underwater scenes, portraits that serve as memorials to locals lost to violence. From the top, get a bird's-eye view of the paintbox-coloured houses lining the slopes beneath. You can go it alone, but better to join in a locally run tour for a deeper understanding of the neighbourhood's transformation.

SEE IT! ***Walk 1km (0.6 miles) west from the San Javier metro station or via 225i bus. Arrange tours at Casa Kolacho near San Javier.***

335

See gorillas in the mist atop Volcanoes National Park

RWANDA // Nothing compares to gazing into the eyes of a gorilla and seeing them look back at you with the same sense of wonder. And there's nowhere better to embrace this experience than in Rwanda's Volcanoes National Park.

The park's stature is impressive, with five steep-sided conical volcanoes, each covered in rainforest, rising out of the East African plains. Their rugged nature becomes apparent as you climb through the thick bush to one of the 10 habituated family groups that live on the Virunga volcanoes' slopes.

SEE IT! *Musanze (Ruhengeri), two hours' drive from Kigali, is the park's gateway.*

Discover the medieval marvel of Great Zimbabwe

ZIMBABWE // Clamber over ancient boulders and ruins, explore narrow crevices and ponder the significance of the greatest medieval city in sub-Saharan Africa, Great Zimbabwe. Dating to the 11th century, this World Heritage-listed site was likely built by ancestors of the modern Shona people as a monarch's palace, surrounded by high, mortarless stone walls. It's best visited at sunrise or sunset when the light gives an other-worldly feel to the landscape.

SEE IT! *Catch a Kombi van taxi from the town of Masvingo to the Great Zimbabwe Hotel; it's a 10-minute walk to the main gate.*

337

Track wildlife in Parc National Odzala-Kokoua

REPUBLIC OF CONGO // Wade beneath the thick canopy of the world's second-largest tropical rainforest in search of Central African megafauna. Lurking in this jungle's depths are hundreds of wild species, including forest elephants and some 22,000 western lowland gorillas.

Now run by NGO African Parks, the park has improved tourism facilities, including viewing platforms, upgraded accommodation and transport links with Brazzaville. Importantly, sustainable projects have also been initiated to provide new incomes for nearby communities to reduce the economic pressures that can lead to poaching.

SEE IT! *Ground transport or flights to the park from Brazzaville are arranged as part of your accommodation.*

© Jakub Czajkowski / Shutterstock

© Paul & Paveena Mckenzie / Getty Images

338

Watch sunset over the Gobi dunes of Mongolia's Khongoryn Els

MONGOLIA // About twice as high and four times as long as the Sahara's famous Erg Chebbi dunes, the Khongoryn Els range in Mongolia's Gobi Desert features some of the largest and most spectacular sand dunes in the world. At up to 300m (985ft), the climb is an exhausting slog that can take the best of an hour, but the views of the desert from the sandy summit are wonderful. Before you lies an undulating blanket of pillow-soft mounds, which intensify in colour from mellow yellow to burnished gold as the day goes on. Locally they are known as Duut Mankhan – the Singing Dunes, a nod to the lilting sound they make when the sand is moved by the wind or as it collapses in small avalanches. Magically, as if from nowhere, locals will appear leading camels to whisk you away.

☛ SEE IT! *Khongoryn Els is in southern Mongolia's Gurvan Saikhan National Park. A mini naadam (traditional games festival) featuring horse racing, archery and wrestling is held here annually in August.*

339

Take a river cruise through Tanjung Puting, an orangutan kingdom

INDONESIA // Travelling by houseboat up a winding jungle river to visit free-roaming orangutans sounds like something out of a child's storybook. Yet it's something travellers do every day in southern Kalimantan, the world's best place to encounter Asia's largest ape. By day, a visit to Tanjung Puting National Park involves lounging on the deck of a *klotok* – a ramshackle, multi-storey liveaboard – your binoculars at the ready, the boat cruising the channel and making impromptu wildlife stops and visits to orangutan-feeding stations. You can also spot unusual proboscis monkeys, toothy crocodiles and swooping hornbills. By night, you'll play games, listen to the sounds of the jungle, count fireflies and retire to comfortable quarters. The journey lasts just three days, but travellers reminisce about it for the rest of their lives.

☛ SEE IT! *Reach the park via direct flights from Jakarta and Surabaya. Boats travel up the Sungai Sekonyer from Kumai to Camp Leakey.*

340

Visit a cathedral of modern art at Tate Modern

ENGLAND // It's been decked with swings, covered in fake sunflower seeds, rent by a great crack and bathed in the light of a strange sun. Tate Modern's Turbine Hall – an impressive space that drops into the earth, its vast roof and industrial concrete making it a perfect secular church – has hosted all manner of showstopping exhibitions. And it's just one space in an art gallery that's full of them.

Tate Modern occupies a former power station on the South Bank of the Thames – it opened in 2000, but such is its effect that it feels like it's been there forever. And it's kept expanding, its permanent galleries (home to a fine collection of post-1900 British art) and main temporary exhibition spaces joined in recent years by a whole new wing, lending this already huge collection a rush of new rooms and an enormously popular viewing gallery.

SEE IT! *The most scenic approach? Either cross Millennium Bridge from St Paul's or take the boat from Tate Britain. It's quietest early on.*

341

Spot puffin chicks and grottoes at Vestmanna Bird Cliffs

FAROE ISLANDS // The surly weather on these rugged north Atlantic islands is no deterrent for the thousands of hardy birds that nest here in summer. If the wind doesn't take your breath away on a boat tour to the Vestmanna Bird Cliffs of northwestern Streymoy, then the sight and sound of screeching sea birds circling and swooping overhead certainly will. Bobbing under soaring cliff-faces, sailing through narrow channels and past mysterious grottoes, visitors can see nesting sites all over the otherwise barren rock, the temporary home for birds such as kittiwakes, guillemots, razorbills and fulmars while they rear their chicks. But the star of the show is undoubtedly the handsome Atlantic puffin, with its bright orange beak, oversized feet and endearingly comical waddle.

SEE IT! *The Faroe Islands can be reached by air, or by ferry from Denmark. Boat trips to the cliffs are very popular; book in advance.*

© Nicola Kota / Alamy Stock Photo

© cdrin / Shutterstock

342

Discover a once-great culture at the Thracian Tomb of Sveshtari

BULGARIA // The Thracians once inhabited large areas of the portion of southeastern Europe now occupied by Bulgaria, and this tomb in remote Razgrad province is one of the most wondrous testimonies to their civilisation. The highlight of the Unesco-protected site is a near-perfectly preserved three-sided tomb full of fine ornamentation, in which elaborate burial gifts were laid. Particularly impressive – and slightly unnerving – are the ten female figures found near the top of the tomb's walls and seemingly supporting the roof, carved intricately but with vacant, spellbound expressions. Despite dating from the 3rd century BC, the tomb was only discovered in 1982.

SEE IT! ***Located 379km (235 miles) northeast of Sofia, the tomb is best reached by car. Isperih, 8km (5 miles) away, has infrequent buses and trains, or take the train to Ruse (90km/56 miles away) then a taxi.***

343

Comb through Seattle's charismatic soul at Pike Place Market

USA // Flying fish, classical music maestros busking for beer money, a street entertainer strumming a guitar while twirling hula-hoops and coffee tourists queueing for fuel outside the world's oldest Starbucks: Pike Place is not your run-of-the-mill market. If Seattle has an essential sight, this is it, a rambunctious collection of permanent stalls where local etiquette invites you to 'meet the producer', slide into Seattle's soul and eat your way from breakfast to dinner without ever touching a fork. As much a symbol of the city as the emblematic Space Needle, Pike Place has been nourishing and entertaining local Seattleites since 1907. Dive in and join them.

SEE IT! ***The market is close to the waterfront in downtown Seattle and is replete with an abundance of cheap food stalls, bars and restaurants. Arrive hungry!***

344

Slim enough to cross using paddle power alone, the aptly named 'narrow fjord' can also be explored by electric ferry

Paddle across Norway's narrow Nærøyfjord

NORWAY // Fjords don't get much narrower or more arrestingly lovely than Nærøyfjord, a steel-blue 17km (11 miles) sliver of water cutting its way through sheer 1200m-high (3930ft) cliffs. Forking south from the main course of Sognefjorden, the Unesco World Heritage fjord measures just 250m (820ft) across at its narrowest point, allowing you close-up views of its breathtakingly steep mountains, hanging valleys and brightly painted wooden clapboard houses. Meltwater makes for impressive waterfalls in spring, but the fjord is just as beautiful in winter when it's all snow and moody monochromes.

You can boat across it by ecofriendly electric ferry from Flåm or Gudvangen. Or for a more peaceful, intimate experience, join a guided half-day kayaking tour in Flåm and look out for the porpoises, otters, dolphins and seals that splash in these waters.

SEE IT! *Departing year-round, two-hour boat trips take in the Nærøyfjord link Flåm and Gudvangen in southwestern Norway.*

© doleesi / Shutterstock

© Marcin Szymczak / Shutterstock

345

Chill out amid colonial splendour at Plaza Mayor, Trinidad

CUBA // The resplendent heart of the colonial Cuban city of Trinidad, Plaza Mayor is an enchanting collection of churches, museums, civic buildings and palm trees that positively oozes centuries-old charm. Anywhere else, this elegant square would be crammed with traffic, but in Trinidad, more sedate rhythms play, though you may spot the odd vintage American jalopy transporting newly married couples from the surrounding churches. The showpiece museum is the Museo Histórico Municipal, a mansion decked out in grand neoclassical style with a perfect view of the city from atop its tower. Take a seat on a park bench and watch the city go by; you might even get an invitation to dance at the energetic socials that fill the square after dark.

SEE IT! *Plaza Mayor is the heart of Trinidad, so you can walk around at your leisure; come at first light for photos or sunset to dance.*

346

Revel in the glory of sumptuous Vank Cathedral

IRAN // Don't be fooled by Vank Cathedral's bland exterior: inside, every inch is covered with rich frescoes and gilded ornamentation, and the eye doesn't know what to admire first. Built between 1648 and 1655 during the reign of Shah Abbas I, the cathedral makes for a dramatic Christian counterpart to Esfahan's tile-tastic Naqsh-e Jahan Square. It cleverly combines church architecture with Islamic design, including a dome-topped sanctuary and warm floral decoration. The masterful artworks depict Bible stories, gruesome martyrdoms and pantomime demons. In the same complex, an interesting museum houses illustrated gospels and Bibles, some from the 10th century. The most intriguing item is a prayer written on a single hair, visible only with a microscope.

SEE IT! *Vank Cathedral is located in the fashionable area of Jolfa, which has been Esfahan's Armenian quarter since the 17th century.*

© Edward Parker / Alamy Stock Photo

© Adrian Studer / Getty Images

347

Spot rare monkeys, manatees and toucans in Mamirauá reserve

BRAZIL // The rare white uakari monkey has a white coat and a bulbous red face, and is found nowhere else on Earth. It's one of the reasons the Mamirauá Sustainable Development Reserve is here. The reserve encompasses moist forest and *várzea*, a forest ecosystem of floodplains and seasonal mud islands. Come here to spot the uakari, as well as jungle critters including toucans, sloths, coatis and anteaters. Follow an indigenous guide on walks along the boggy riverbanks, binoculars bobbing at your neck, or canoe the flooded forest, crossing your fingers for a manatee sighting. In June and July, the extremely lucky might even glimpse a jaguar slinking by. Stay at the Pousada Uacari ecolodge, built on floating logs – sustainable travel at its best.

SEE IT! ***Travel by boat into the reserve from the access town of Tefé; your accommodation will book your transfer.***

348

Court the mythical landscapes of the Fitz Roy Range

ARGENTINA // With its rugged wilderness and shark-toothed summits, the Fitz Roy Range is the de facto trekking capital of Argentina. While world-class climbers take on the toothy summits of Cerro Fitz Roy and Cerro Torre, mere mortals content themselves trekking through moss-clad lenga forest and visiting alpine tarns. The northern sector of Parque Nacional Los Glaciares resembles Chile's Torres del Paine – it's close as the condor flies – but with exceptional access to village life, with steakhouses, microbrews and warm beds in the frontier town of El Chaltén awaiting hikers. Yet it never stops being Patagonia, so even a summer's day under bluebird skies could bring bone-chilling winds. But nobody would say it's not worth it.

SEE IT! ***The gateway town of El Chaltén is accessed by bus from El Calafate. Avoid winter (May to September), when facilities shut down.***

349

Walk beneath whales at the Natural History Museum

ENGLAND // Eighty million specimens, billions of years – London's Natural History Museum squeezes eons into its formidable space. Here you can learn about dinosaur bones and butterflies, polar bears and human babies, volcanoes and the Big Bang. This den of knowledge, treasures and research is housed in a majestic 1881 building. When the famous replica diplodocus in the entry hall was replaced by a gorgeous, curving blue whale skeleton, it felt like a fitting change. With its ancient rocks and prehistoric mammals, this spectacular museum celebrates the past – but in an age of extinction, it's also a desperate love letter to the present.

☛ SEE IT! *It's five minutes' walk from South Kensington Tube.*

350

Dive through psychedelic forests on Rainbow Reef

FIJI // Divers are spoilt for choice in Fiji's archipelago, where waters have been dubbed the soft-coral capital of the world. Strong tidal currents push the deep water back and forth through the Somosomo Strait, providing such rich nutrients for the soft corals and sea fans that they form psychedelic underwater forests. Highlights include an immense drop-off covered in luminescent corals known as the Great White Wall, and Annie's Brommies – an outcrop teeming with fish.

☛ SEE IT! *Flights service Taveuni, from where the reef is easily accessed. Avoid January and February.*

351

Discover Parc National d'Isalo's desert playground

MADAGASCAR // This outlandishly beautiful national park may be a desert canyon, but get your bathing costume ready to dive right in. Between explorations of vertical walls, chiselled buttes and canyon floors by foot, mountain bike, horse or even *via ferrata* (cabled climbing routes), it's hard to resist the streams meandering beneath leafy canopies. Come day's end, you'll sit happily exhausted and watch the landscape shift with the golden rays and shadows cast by the last light.

☛ SEE IT! *Taxis-brousses pass the park while travelling between Tuléar and Antananarivo – grab an empty seat.*

© Justin Foulkes / Lonely Planet

352

Roam the empire's wild frontier at Hadrian's Wall

ENGLAND // Walking the Hadrian's Wall Path, a delightfully wild 135km-long (84-mile) trail running coast-to-coast across England, you'll appreciate how desperate the Romans were to stamp their authority on northern Britain. Built between AD 122 and 128, and named after the emperor who ordered its construction, Hadrian's Wall was an amazing feat of engineering, even by Roman standards. It stretches from Wallsend in the east to Bowness-on-Solway in the west. After each Roman mile (roughly 0.9 of a modern mile) a guarded gateway called a milecastle was built, each meticulously numbered. The remains of several of these can still be found along the wall, as well as several forts and a total of six museums that explain its history. Housesteads is a well-preserved fort with great views and a museum, while Segedunum has a fascinating exhibition exploring daily life on the wall.

SEE IT! ***Hiking the wall takes five to 10 days. Alternatively, visit the likes of Housesteads (near Hexham) and Segedunum (in Wallsend).***

© leoks / Shutterstock

353

Uncover a Scandi time capsule in Gamla Stan

SWEDEN // History oozes from the edifices in Stockholm's old town. Here, cobblestone streets wriggle past Renaissance churches, baroque palaces and medieval squares, while spice-coloured buildings sag like wizened old men. Founded in 1252, Gamla Stan has been blackened by plague and famine, consumed by flames when the castle of Tre Kronor was burned to the ground, and embattled by Danish and Swedish factions. The innocent-looking square of Stortorget was the scene of a bloodbath in 1520 when a Danish king tricked, trapped and beheaded 82 rebellious Swedes. Dominating the skyline and teetering above Gamla Stan, Storkyrkan is the old town's medieval cathedral and one-time royal coronation venue; inside you'll find a dramatic 15th-century depiction of St George, wrestling his legendary dragon. Nearby, Kungliga Slottet is the official residence of the King of Sweden.

SEE IT! ***Västerlånggatan is Gamla Stan's nerve centre, a bustling thoroughfare lined with galleries, restaurants and shops.***

Opposite, from top: Valencia's City of Arts and Sciences is a fabulous celebration of modern design; away from the madding crowds at Great Barrier Island

354

Go high-tech at the eye-popping City of Arts & Sciences

SPAIN // Poor old Valencia never had an Alhambra or Sagrada Família to match its urban amigos in Granada and Barcelona. Rather than sitting around and sulking, city planners went away and designed their own super-modern monument for the ages, the spectacular City of Arts & Sciences. Built on a reclaimed riverbed in the 1990s using the futuristic designs of local boy Santiago Calatrava, the project proved to be controversial, long-winded and expensive. The finished article, with its glassy, curvy, futuristic flourishes, accommodates an opera house, a science museum, a 3D cinema and an aquarium. Kicking off the 21st century like a Messi wonder-goal, it has every right to claim its place alongside the Alhambra et al as a modern Spanish masterpiece.

SEE IT! *A dozen high-speed AVE trains run between Madrid and Valencia daily. Combined tickets for the site's venues are sold online.*

355

Head off-grid on Great Barrier Island

NEW ZEALAND // In the outer Hauraki Gulf, mountainous Great Barrier Island is New Zealand 100 – or 1000 – years ago. Think dirt roads, ramshackle shops selling dusty tinned food, and – gasp – poor or no cellphone service. You come here for hikes through the hushed kauri forests, kayak trips along the craggy coastline, soaks in natural rock pools and dazzling stargazing – the island is an International Dark Sky Sanctuary. The native bush is thick, thanks to a lack of invasive species; there's hardly anything more glorious than a stroll along the mountain track at sunset, the light turning the ridges and valleys gold and pink. Surfing is decent and diving is excellent – visibility reaches 33m (108ft) on a good day.

SEE IT! *From Auckland, take a 30-minute flight or a five-hour ferry and then rent a car. December is busy season – book ahead.*

354

355

Meet a Renaissance icon at the Galleria dell'Accademia

ITALY // He's been struck by lightning, attacked by rioters and had his toes bashed with a lunatic's hammer, so waiting a few minutes for your date with *David* is a minor inconvenience. At first sight, you'll be seduced by his nonchalant stance and the beauty of his poised body, carved by Michelangelo out of Carrara marble in 1504. Thick, luscious curls frame his lovely face, which looks serene and boyish from the left. But, when you lock eyes, you'll feel the terrible intensity of his gaze as he meets his own date with destiny to bring down Goliath.

☛ SEE IT! ***It's 15 minutes' walk from Florence's Stazione di Santa Maria Novella. Book at www.firenzemusei.it to reduce queuing time.***

Contemplate the tragedy of war in Flanders Fields

BELGIUM // Over four blood-soaked years along the Western Front during WWI, almost a whole generation of young men perished amid the mud, barbed wire and trenches. There's no single place called Flanders Fields – the name comes from a haunting poem by John McCrae – but the war cemeteries and memorials that straddle the French-Belgian border stand as silent witnesses to the deaths here. As well as statues, arches and endless avenues of neat white crosses, red poppies appear across the countryside every summer like the ghosts of the fallen.

☛ SEE IT! ***Ypres is the gateway. The city's Menin Gate is engraved with the names of 54,896 soldiers whose bodies were never found.***

Pick your desert-island dream in the Exuma Cays

THE BAHAMAS // If, as the cliche has it, the Caribbean islands are like a string of jewels set in a turquoise sea, then the Exumas have been cut and polished by the most skilful hand. A collection of more than 300 islands and cays, this is a destination made for pottering about in tiny sailboats or paddleboards on blissful beaches and reef-fringed coastlines. George Town has colonial ruins, and there are nature walks on Stocking Island, but most visitors will be quick to grab a snorkel to explore the islands' extraordinarily rich sea life.

☛ SEE IT! ***You can fly from Nassau to George Town on Great Exuma, but you'll adopt the island vibe quicker if you take the ferry.***

Visit Kilwa Kisiwani, East Africa's grandest medieval city-state

TANZANIA // In its 13th-century heyday, Kilwa Kisiwani ('Kilwa on the Island') was a seat of sultans and centre of a trade network linking old Shona kingdoms and the gold fields of Zimbabwe with Persia, India and China. It was renowned far and wide, with the chronicler Ibn Battuta describing its exceptional beauty. Today, Kilwa's substantial and evocative ruins – including palaces and mosques – whisper of bygone days of glory and offer testimony to the influence of Swahili culture. Several buildings have been restored, with informative signboards.

☛ SEE IT! ***The ruins are a short boat trip from Kilwa Masoko on Tanzania's southeast coast. Visit early; guides are compulsory.***

360

Nod off on a sub-zero sleepover in the original Icehotel

SWEDEN // Staying at Swedish Lapland's Icehotel, which is carved afresh each winter from the frozen Torne River, 200km (125 miles) north of the Arctic Circle, is a like stepping into Narnia, *Frozen* and *Sleeping Beauty* combined. Surprisingly comfortable ice beds topped with reindeer skins and expedition-grade sleeping bags will keep you warm despite the -5°C chill. But enchantment of lying in a glistening blue-white cocoon, snow-muffled and sparkling all night, may just keep you awake.

SEE IT! *Jukkasjärvi is 14km (7 miles) from Kiruna Airport. Transfers from the airport to the Icehotel by husky-sled are available.*

361

Trek the USA's only tropical forest at El Yunque rainforest

PUERTO RICO // Puerto Rico's premier entry on this list will not surprise those well acquainted with this Caribbean US commonwealth, but might shock the uninitiated: above any of the island's fabled beaches comes the inland wonder of El Yunque, a waterfall-riven hunk of rainforest containing a whopping 1000-plus plant species in a 113-sq-km (44-sq-mile) tropical enclave. Exploring is made easy by well-maintained trails and visitor facilities, including some swimming holes.

SEE IT! *You can visit from San Juan, 45km (28 miles) southeast, or stay over at atmospheric accommodation, such as Rainforest Inn.*

360

© Wolfgang Kaehler / Getty Images

362

Feel Amarbayasgalant Khiid's untouched spirituality

MONGOLIA // Surrounded by gentle green hills, Amarbayasgalant Khiid is one of the most important sacred sites in Tibetan Buddhism. Built in the 1730s by the Manchu emperor and dedicated to Zanabazar – Mongolia's first Gelugpa (Yellow Hat) spiritual leader – this remote monastery is the country's best preserved. It languished (but was mercifully spared) under Soviet rule. With its north-south layout and guardian deities, there is a strong Manchu influence, but the site's gentle decay and peaceful remoteness give it an undeniably special feel.

SEE IT! *It's north of the town of Selenge in northern Mongolia – travellers need 4WD hire or their own transport to get around.*

© krzych-34 / Getty Images

363

Sip wine in the cavernous cellars of Mileştii Mici

MOLDOVA // Looking for the world's largest wine collection? Descend into the mind-bendingly massive underground wine cellars at Mileştii Mici, which sprawl for about 200km (124 miles) and stock around 1.5 million bottles of the stuff.

A wine-tasting trip here is an experience you'll never forget: in order to navigate the huge cave network, tours are done by car, while the impressive tasting room is hidden behind a 3-tonne secret stone door.

SEE IT! *The winery is 20km (12 miles) south of Chişinău. Book a tour (including car, driver, tastings and lunch) through a travel agency.*

364

Discover the multilayered history of the Château de Chenonceau

FRANCE // The turrets, the river-spanning archways, the impeccable formal gardens... to describe this Loire Valley castle as grand is an understatement. The Château de Chenonceau dates from the 16th century, and many a wild party was thrown here by a succession of aristocrats, including Catherine de Médicis. The *pièce de résistance* is its window-lined Grande Gallerie spanning the Cher River. During WWII, the Cher was the boundary between free and occupied France, and the gallery was used by escaping Resistance members and refugees.

SEE IT! *From Chenonceaux (spelt with an 'x'), the town just outside the château grounds, frequent trains run to the city of Tours.*

© Don Mammoser / Shutterstock

© Perfect Lazybones / Shutterstock

365

Walk through paradise along the Lavena Coastal Walk

FIJI // The most celebrated of Fiji's lush walking trails, this 5km (3-mile) coastal walk has all the components of the perfect wander in paradise. It follows the rainforest edge along stunning white-sand Lavena beach, passes peaceful palm-thatched villages, crosses a rickety suspension bridge and traverses the ancient valley of Wainibau Creek, with scenery straight out of Jurassic Park. Just as you think the tropical heat is becoming unbearable, the finale: a gushing waterfall. That first dive into the cool, deep waters feels like a piece of heaven.

SEE IT! *Lavena is on Taveuni, connected to Fiji's main island via flights to Nadi and Suva. In November to April floods can close the falls.*

366

See the organic art of the Ifugao (Banaue) Rice Terraces

PHILIPPINES // Ancient, yet still used today. Rudimentary, yet sublime. The emerald Ifugao rice terraces are impressive for many reasons. These mud-walled shelves were created more than 2000 years ago by the once-headhunting Ifugao people in an amazing manipulation of nature. The shelves have been impossibly carved into sharply sloping mountains and undulate for miles, forming a dizzying spectacle. The terraces are most picturesque one to two months before rice harvest, when they become bright green before gradually turning golden.

SEE IT! *Banaue is nine hours from Manila by bus. Go June to July (before harvest) or February to March (cleaning and planting).*

© Oleg Bakhirev / Shutterstock

367

Revel in the cornucopia of colour at Monet's garden

FRANCE // No matter your impression of impressionism, you can't fail to be moved by the loveliness of Claude Monet's pink-hued house and flower-filled gardens at the Maison et Jardins de Claude Monet in Giverny, which he planted – and painted – while living here for his final 43 years. You'll recognise settings from some of the artist's most famous works, including his *Japanese Bridge* and water-lily-filled *Jardin d'Eau* (Water Garden). From early to late spring, daffodils, tulips, rhododendrons, wisteria and irises appear, followed by poppies and lilies. By June, nasturtiums, roses and sweet peas are in flower. Come September, the gardens are a riot of hollyhocks, dahlias and sunflowers.

SEE IT! ***The nearest train station is 7km (4 miles) to the west at Vernon, from where seasonal shuttle buses, taxis and cycling/walking tracks run to Giverny. It's closed November to Easter.***

368

Take the prehistoric path to Boiling Lake

DOMINICA // Did the devil forget to plug a hole to the underworld? It certainly looks that way in Dominica's Valley of Desolation, where a billowing plume of vapour signals the approach to Boiling Lake. The gruelling 6km (4-mile) hike to reach it, through the Jurassic-looking folds of Morne Trois Pitons National Park, ranks among the Caribbean's most pinch-me wilderness adventures. Crumpled green hills and volcanic peaks dictate the trail, which eventually leads to the 610m-wide (2000ft) blue-grey flooded fumarole – a crack in the earth that allows hot gases to vent from the molten lava below, and the world's second-largest hot lake (after New Zealand's Frying Pan Lake).

SEE IT! ***The hike starts from Ti Tou Gorge in southern central Dominica, and a guide is recommended (check the tourist board website for recommended tour operators).***

369

Ride the rails of the Ffestiniog & Welsh Highland Railway

WALES // There is a sedate, stylish way to see Snowdonia, North Wales' mountainous massif, and this is it: by a vintage steam rail ride festooned with blockbuster scenery. Watch the beauty spots unfold from Snowdon itself, to picturesque Beddgelert and mighty Caernarfon Castle.

SEE IT! ***Start the ride at Blaenau Ffestiniog, a former slate-mining town transformed into an adventure magnet.***

370

© Westend61 / Getty Images

370

Discover El Jem, Africa's great Roman amphitheatre

TUNISIA // What did the Romans ever do for us? Well, at El Jem, they built one of the wonders of North Africa. In its 3rd-century heyday, this great coliseum could accommodate 35,000 spectators, who bayed for blood as gladiators pitted muscle and metal against wild animals and each other for the pleasure of the emperor. This was one of the largest amphitheatres in the Roman world, and the biggest in the Africa dominion, and it survives largely intact. Unlike the Colosseum in Rome, the theatre at El Jem soars above the surrounding city. In the central arena, you can still imagine the surge of adrenaline as wild beasts were released. If any of this looks familiar, El Jem has starred on the big screen, including a cameo in Monty Python's *Life of Brian*.

SEE IT! ***The amphitheatre dominates the centre of the Tunisian town of El Jem, an easy day trip from Sousse or Sfax.***

371

Be terrified by reality at Budapest's House of Terror

HUNGARY // This is no descent into fantastical horror, as this Budapest museum's name might imply, but a hard, detailed look at the very real fascist and socialist regimes (the 'Double Occupation') under which Hungary spent much of the last 75 years. And as a result it's infinitely more chilling. The museum is located in the former headquarters of the dreaded ÁVH secret police, and it is poignant to chart the atrocities that occurred in a building where many were actually perpetrated. As you learn about the detention, torture and execution of the regimes' victims (walls were built thick to muffle screams) you perceive that this is about the Double Occupation's darkest sides only, but there are no other visitor experiences in Eastern Europe that so comprehensively document what life (and death) beyond the Iron Curtain could be like.

SEE IT! ***Take Budapest's M1 metro line to Vörösmarty utca.***

Opposite, from top: Plan a grape escape to the Alto Douro; beat the Saharan heat in an ancient underground bolt-hole

Romancing the vine in Alto Douro wine country

PORTUGAL // Quite possibly the fairest of all Europe's wine regions, the Unesco-listed Alto Douro unfurling east of Porto is divine. You'll come to taste bold reds grown in schist soils and to hone your knowledge of the best port wines in the world by tasting rubies and tawnies and end up lingering longer than planned for the scenery. Ah, the scenery... Here, incredibly steep, terraced vineyards comb the contours of hillsides that spill down to the mighty Douro River.

The Douro is never lovelier than on a gold-kissed autumn day, when grapes hang heavy from the vines and the harvest is in full swing – visitors are, incidentally, welcome to join in. Stay overnight at a quinta (wine estate) for the ultimate experience.

SEE IT! *Arguably the best way to explore is to take the three-hour train ride from Porto's São Bento station to Pocinho.*

Have a *Star Wars*-style sleep in Matmata

TUNISIA // *Star Wars* fans may come just on a film-inspired pilgrimage, but this village of underground pit houses is more than just the stage for Luke Skywalker's home. Because of the intense Saharan heat, the Berbers who live here have long sought shelter by carving ingenious dwellings below the surface. Today, many of the houses can be visited and also function as simple hotels. The most well-known dwelling is the Hotel Sidi Driss, which stood in for the Lars family homestead in the original *Star Wars* and is still littered with bits of the set. Roaming the lunarscape surface, pockmarked by troglodyte trenches, and then spending the night cocooned in your very own cave is the closest you'll come on earth to the experience of sleeping in outer space.

SEE IT! *Shared taxis from Gabès (30 minutes) run to Matmata Nouvelle, from where another connects with old Matmata (15 minutes).*

372

373

© Digital Nature Scotland / Shutterstock

Kengo Kuma's striking building is a fittingly fabulous home for the V&A Dundee, Scotland's newest design museum

374

See museums, polar ships and a city reborn on Dundee Waterfront

SCOTLAND // Is it a ship? Is it a cliff? The flagship V&A Dundee – which takes the form of two enormous slatted, inverted pyramids – is a striking arrival on a waterfront that's seen all manner of sights. At its peak, Dundee built over 200 ships a year, including Scott of the Antarctic's *Discovery*. It saw off whalers and trading vessels stacked with jute and jam. The docks faded, but change is afoot. The V&A Dundee, which houses Scotland's biggest museum space and the best in Scottish design, is only part of the area's reinvention. The £1 billion project, which is due to finish in 2031, has been controversial. But gone are tangled roads and gloomy municipal buildings, replaced by gleaming modern structures, a park and a new station. *Discovery Point* (Scott's ship returned to its home port in 1986) is another must-visit, a triumph of grace and toughness designed to resist temperatures of -30°C. The redevelopment shines a spotlight on a city with fine museums and a glorious natural setting, where land meets sea and the thrum of Scotland's central belt bleeds into the beauty of the Highlands.

SEE IT! ***The train station is moments from V&A Dundee, with regular services to Edinburgh, Glasgow and London.***

© Ventura / Shutterstock

© f11photo / Shutterstock

375

Feel the wild Atlantic force at Cabo de São Vicente

PORTUGAL // Until the late 14th century, the lonely, wind-battered headland of Cabo de São Vicente at Europe's southwesternmost tip was the end of the world. Then along came the bold Prince Henry the Navigator, with an aching desire to set sail to far-off countries in search of sugar, spice and all things nice. The high cliffs of this cape would have been the last thing he glimpsed as he caught the breeze and drifted towards the unknown in the storm-tossed Atlantic. With wide-open horizons, crashing waves, a blinking lighthouse and the occasional fisher precariously dangling a line from the cliffs, the cape still feels wild and remote today. Come to see the sinking sun silhouette the headland that played such a pivotal role in maritime history.

SEE IT! ***Cabo de São Vicente is at the southwest tip of Portugal's Algarve. The closest town is Sagres, 7km (4 miles) east.***

376

Behold America's birthplace at Independence National Historic Park

USA // It would be unthinkable to come to Philadelphia without paying homage to America's creation legend in this shady city park, dotted with historic buildings where the seeds for the Revolutionary War were planted and where the US government came into being. As you cross the threshold of Independence Hall and stand in the very room where the founding fathers signed the Declaration of Independence in 1776, it's hard not to be awestruck. Same when you clap eyes on the whopping Liberty Bell that tolled America's first note of freedom. George Washington, Thomas Jefferson, Ben Franklin, Alexander Hamilton – all of the stars of US history were here at the get-go to help the fledgling nation get its democratic ideals on.

SEE IT! ***Take the subway train to 5th Street Station, then hit the visitor centre to pick up timed entry tickets for Independence Hall.***

© Hieronymus Ukkel / Shutterstock

© Viktoriia Ablohina / Shutterstock

377

Get into Hamburg's groove at the Elbphilharmonie

GERMANY // No building is more emblematic of Hamburg's architectural zeitgeist than the Elbphilharmonie. Designed by Pritzker Prize–winning Swiss architects Herzog & de Meuron, its curving glass panels rise up like the crest of a great wave to break above a 1960s redbrick warehouse that was once used to store cocoa beans.

The approach is spectacular via Europe's longest escalator to the Plaza, where a wraparound balcony commands far-reaching vistas. Book tickets for a performance in the acoustically magnificent Grand Hall or, if you'd rather just kick back with dress-circle views, head up to the 12th-floor Harbour Bar for drinks as the sun goes down over HafenCity.

☛ SEE IT! *The Elbphilharmonie is the most visible icon of the harbourside HafenCity district, a shining model of urban regeneration. Guided tours (some in English) are listed on its website.*

378

Trace the tortured artist's life at the Van Gogh Museum

NETHERLANDS // He epitomised the artist's struggle through poverty and obscurity – only selling one painting in his lifetime – but Vincent van Gogh was the greatest 19th-century Dutch painter. Housing the world's largest collection of his works, this Amsterdam museum features more than 200 canvases, from his dark, sombre-themed early career in the Netherlands to his later years in sunny France, where he produced his best-known work with its characteristic giddy colour. You'll see the evolution from floor to floor as you marvel over vivid yellows, deep purples and rich blues in renowned paintings such as *Sunflowers* and *The Yellow House*. Also here are 400 of his drawings and 700 letters.

☛ SEE IT! *Located in the Old South neighbourhood, the museum is easily reached by tram. Purchase tickets online as far in advance as possible.*

379

See the sun set in sloping pools at Longji Rice Terraces

CHINA // For hundreds of years, these 800m-high (2625ft) terraced paddy fields remained unknown to travellers, then everything changed in the 1990s when a photographer named Li Yashi moved here. His amazing images put mesmerising Longji (literally 'Dragon's Back') firmly on the tourist trail. A feat of farm engineering, the fields cascade in swirls that resemble the contours of a giant thumbprint, sometimes immersed in a sea of cloud. The hills are dotted with centuries-old minority villages, which you can hike between, or there's a cable car connecting the principal viewing points.

SEE IT! *Guilin is the gateway city. Ping'an, Dazhai and Tiantouzhai villages offer the most spectacular views.*

John Gregory Smith's Top Five Places

British chef, author and presenter John Gregory Smith specialises in Middle Eastern cuisine. He has written six cookbooks. His latest, Saffron in the Souks, *is out now.*

01

FAROE ISLANDS – This remote archipelago in the North Atlantic has some of the most dramatic hiking in the world. Filled with rugged cliff tops, cascading waterfalls and misty valleys, it's the perfect place to get away from it all.

02

MACHANE YEHUDA MARKET, JERUSALEM – Packed with hipster coffee shops, Iraqi kebabs and sticky sweet *rugelach*, Friday is the day to explore this food market. Go for the food and stay for the party vibes.

03

ANTIPAXOS, GREECE – Dive off a boat into the azure waters that surround this tiny island off Corfu and enjoy listening to nothing but the waves lapping the deserted shoreline. Bella Vista does a cracking plate of grilled calamari.

04

SOHO HOUSE, GREEK STREET, LONDON – My favourite bar in the world and the place I like to unwind when I am home in London. I favour a pepperoni pizza and an espresso martini or two. It's open late so I can have a good old run around until the early hours.

05

QUADISHA VALLEY, LEBANON – This massive valley is scored deep into the mountains. Villages cling to the clifftops and there are hundreds of monasteries, one of them run by a hermit related to Pablo Escobar.

© JAZZDOG / Shutterstock

Art and artefacts sit side by side in the eclectic, beautifully lit galleries of the MONA museum

380

Admire art at the end of the world at MONA

AUSTRALIA // Watch dolphins play as you arrive by motorboat to Tasmania's fabulously eccentric Museum of Old and New Art (MONA), built into a cliff on the River Derwent near Hobart. Commissioned by millionaire professional gambler David Walsh as a way to give back to his home city, MONA opened to much acclaim in 2011. Visitors wander the maze-like underground space with headphones, which replace traditional museum labels. Exhibits range from ancient Egyptian sarcophagi and Stone Age pottery to avant-garde sound-and-light sculptures, eclectically positioned without regard to time period or genre. Outdoors, sip wine from the surrounding vineyard while gawking at enormous installation pieces. You can even sleep in ultra-modern riverside lodges decorated with art from the MONA collection.

SEE IT! *MONA can be reached by boat, bicycle or bus from Hobart. Fly to the city from most other Australian cities.*

Forget the noisy 4WDs – hop in a canoe for a serene safari that offers a low-impact intro to Zimbabwe's wildlife

Explore the wild, watery world of Lower Zambezi National Park

ZAMBIA // Wilderness, water and dramatic scenery are the defining elements of a safari in Lower Zambezi National Park. The park, part of a transfrontier conservation area also encompassing Zimbabwe's Mana Pools National Park, is centred on a wildlife-rich floodplain spreading out from the Zambezi, one of Africa's mightiest rivers. Parts are densely vegetated, with mopane woodland alternating with acacias, and stately ebony, baobab and fig trees near the riverbanks.

Apart from its natural wildness, one of Lower Zambezi's main attractions is water-based safaris. Paddle silently in a canoe past splashing elephants, half-submerged hippos and slumbering Nile crocodiles. Float past zebras and buffaloes while marvelling at more than 400 species of birds. Whether you choose a canoe or a larger boat, riverine safaris offer a serene new perspective, without the disturbing thrumming of a 4WD. Most lodges are situated directly on the riverside, which allows you to continue to appreciate the water's rhythms once back on land. Expertly guided walking safaris can be arranged, as can hikes up the steep Zambezi escarpment, with views over the river valley and into Zimbabwe.

SEE IT! *Get there by public transport plus a boat ride, by vehicle or by light aircraft from Lusaka, Livingstone or Mfuwe. The best time to visit is May to October.*

© Philip Lee Harvey / Lonely Planet

© Douglas Lander / Alamy Stock Photo

382

Devour forensically detailed displays on the fall and rise of Sweden's best-known warship

Admire a mighty nautical miscalculation at Vasamuseet

SWEDEN // When the mighty Swedish warship *Vasa* sank to the bottom of Stockholm harbour in 1628 – just 1300m (4265ft) into her maiden voyage – it was an embarrassment to the Swedish crown. Built with devastating power in mind, *Vasa*'s potential was never realised, a grave design flaw sealing her fate. She lay on the seabed for 300 years before being carefully raised and painstakingly restored. In what must be one of the greatest comebacks of all time, *Vasa* is now a source of pride and a symbol of the Swedish Empire at its pinnacle. On display in full regalia, the ship is glorious indeed, and a trip to Scandinavia's most-visited museum is worth every krona. Mind-bendingly intricate displays over five levels detail the events of the sinking and the salvage, as well as the ongoing efforts to preserve this piece of Swedish history. You can even step aboard a reconstruction of the upper gun deck to feel what it might have been like to be on the ship. Vasamuseet would almost be too much to take in if, unlike its subject, it wasn't so well designed.

SEE IT! ***Vasamuseet is on the island of Djurgården in Stockholm. From June to August arrive early to avoid summer crowds.***

© Tomas Zrna / Getty Images

© Maykova Galina / Shutterstock

383

Take a time-out within the desert of Sharqiya Sands

OMAN // Want to get a taste of less-hurried Omani life? Beeline south from Muscat to Sharqiya Sands (referred to locally as Wahiba Sands). This desert with rolling, rippling orange dunes at its core is where Oman's traditional Bedu life clings on. The harsh but beguiling landscape is best appreciated by slowing down rather than revving up, so devote some time rather than opting for a day trip. Overnight at one of the desert camps that are encroached by the sand and eschew the 4WD excursions to explore via a full-day camel safari. Come between Sunday and Wednesday to avoid the weekend-jaunt crew from Muscat.

SEE IT! *A 4WD is necessary to get to the Sharqiya Sands' camps. Some camps provide a transfer service from villages on Hwy 23.*

384

Gaze in awe at a giant shiny statue of Chinggis Khaan

MONGOLIA // The massive silver statue of Mongolia's hero is a bucket-list sight for most visitors to the country. At 40m (130ft) high, this is one huge Chinggis, sat sternly astride his horse at the spot where the real Chinggis purportedly found a golden whip. The statue, built in 2008, manages to earn its way onto myriad 'oddest attraction' lists, simply because of its size and gleaming surface, which can be seen for miles across the steppe. The complex includes an impressive museum with Hunnu artefacts and items from the Mongol empire, and a short film describes how the monument was built.

SEE IT! *The statue is located at Tsonjin Boldog, about 55km (34 miles) east of the capital of Ulaanbaatar and accessible by taxi.*

385

Escape to black beaches and snowy peaks on the Arctic Coast Way

ICELAND // Majestic cliffs. Glacial fjords. Crashing waves. Lounging seals. For a wild adventure in Iceland, head for the Norðurstrandarleið (Arctic Coast Way). The route from Hvammstangi in the west to Bakkafjörður in the east includes six peninsulas stretching north, tracing 900km (560 miles) of Iceland's most rugged, remote coastline. It's your chance to experience heart-pounding natural drama and tranquil escape. From saga settlements to whale-watching hubs, each of its small towns offers special insight into life on the edge of the Arctic.

SEE IT! *The Arctic Coast Way is a journey in the pioneering spirit. Some roads can be very rough – renting a 4WD is recommended.*

© zysman / Getty Images

386

Discover China's creative side at Beijing's 798 Art District

CHINA // Red Maoist slogans and statues of burly communist workers show Beijing's premier art district is loud and proud about its proletarian environs. Set inside the cavernous buildings of a disused electronics factory, the workshop setting is the perfect canvas for ambitious projects requiring lots of space, and the area has become a landmark for China's leading artists. Cafes dot the streets, providing a welcome relief in between gallery hopping: prepare to be alternately bemused and captivated.

SEE IT! *798 is in Chaoyang district; the nearest subway station is Wangjing South. Exploring is easy, thanks to English-language maps.*

© Matt Munro / Lonely Planet

Feel America's welcome at the Statue of Liberty and Ellis Island

USA // Torch held high, Lady Liberty has long greeted newcomers sailing into New York Harbor in search of a better life. The giant statue is the USA's premier icon, and peering out from her crown at New York's skyline is an absolute thrill. Ellis Island floats nearby, the gateway where 12 million immigrants arrived and made America what it is today. Walk in their footsteps, and sense the hope and freedom of starting life anew.

SEE IT! ***Ferries to both sights depart from Battery Park at Manhattan's southern tip. You're best buying tickets in advance.***

388

Hike trails carved through time in Tiger Leaping Gorge

CHINA // Tiger Leaping Gorge is one of nature's prettiest poster children. Snowy mountains rise on either side of a gorge so deep you can be 2km (1.2 miles) above the Jinsha River. You wind up and down trails through tiny farming villages, where you can rest while enjoying glorious views. The knee-wrecking climb might make you whimper at times – but China's unmissable trek is gorgeous every step of the way.

SEE IT! ***The gorge carves its way through remote northwest Yunnan, near the Unesco-listed town of Lijiang.***

Climb the soaring sea cliffs at Sliabh Liag

IRELAND // In a windswept corner of Donegal, the sea cliffs of Sliabh Liag soar some 600m (1970ft) above the waves of the wild Atlantic. A rough path leads along a narrow ridge to the summit, making for a dizzying walk. Rolling mists only add to the romance, while the sheep grazing by the cliff edge seem inclined to pose for pictures. The Cliffs of Moher might be more famous, but these ones are higher – and free!

SEE IT! ***Though most people come by car, local buses stop in the village of Carrick, from where it's possible to hike to the cliffs.***

Find true value at Stirling Castle

SCOTLAND // Stirling Castle's history takes in numerous monarchs, eight sieges and some serious skulduggery, and visiting is a wonderfully vivid experience. Set above the city's Old Town, with a sheer drop on its western side, it mixes formidable battlements with charming gardens, and hidden nooks with bright tapestries. There's plenty for the eyes to feast on, from the joyfully colourful restored Royal Palace to careful archaeological displays and splendid views across the ramparts to the Highlands.

SEE IT! *You can be here in an hour from Edinburgh or Glasgow by train. Visit in late afternoon to feel you've the castle to yourself.*

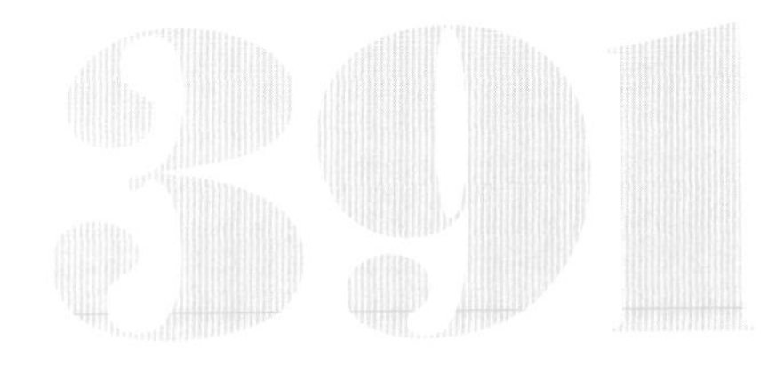

Soak up the Grand Place's majestic guildhalls

BELGIUM // Europe is full of grand squares but there's something special about Brussels' Grand Place (Grote Markt). As you stroll through backstreets crammed with restaurants and boutiques, you might not even realise it's here until you emerge into this awe-inspiring space. Majestic buildings, magically lit up at night, huddle side by side, highlighting the opulent guildhalls' varying architectural styles. It's the beating heart of the city, with buzzing cafes, bloom-filled markets and festivals; from Christmas fairs to an incredible biennial 'flower carpet'.

SEE IT! *Brussels is accessible by Eurostar and other fast trains. From Brussels airport, it's a 20-minute train ride to the city centre.*

Be stirred by Wadden Sea National Park

DENMARK, GERMANY, THE NETHERLANDS // Unesco-anointed for being the world's largest expanse of intertidal sand and mud flats, Wadden Sea National Park sprawls across three countries' territories, and is a habitat for diverse plants and animals. There's much to learn at the park's visitor centres, but take a boat trip to one of the islands, such as Texel in the Netherlands or Denmark's Fanø to experience idyllic coastal settlements. In autumn, witness up to a million starlings taking to the skies to perform a sky-obliterating murmuration.

SEE IT! *The park can be visited from Denmark, Germany or the Netherlands, and Sort Sol is best seen from Ribe in Jutland, Denmark.*

Take in unrivalled coastal beauty at Acadia National Park

USA // The mountains meet the sea in this protected patch of eastern Maine that looks like an artist painted it. Waves crash against pink-coloured cliffs by the waterfront, while interior paths climb through deep-green forest to ridge tops where osprey and eagles fly overhead. The high point is Cadillac Mountain, the 466m (1530ft) peak from which early risers can catch the USA's first sunrise. Scenic ponds, carriage roads, bridges and teahouses fill out the pretty picture.

SEE IT! *Bar Harbor, Maine, is the hub for visits. Go May to October for good weather and open facilities; July and August bring the crowds.*

394

Explore past and present at Royal Albert Dock

ENGLAND // While many of Britain's rebooted industrial sites are compelling places, few can rival the splendid Royal Albert Dock Liverpool. When it opened in 1846, it was cutting edge; built with iron, brick and stone, it used the world's first hydraulic cranes, but it closed in 1972. Now, its chunky red-brick warehouses are home to Liverpool's biggest attractions. The Beatles Story features Fab Four memorabilia. Tate Liverpool packs in modern art and major touring exhibitions. The Merseyside Maritime Museum tells of the *Titanic*, and the nine million people who left via Liverpool for the United States and beyond. And the International Slavery Museum explores the grim trade in which the city was a major player, presenting ships' logs, shackles and the stories of the men and women who wore them.

SEE IT! ***The docks are a 20-minute walk from Liverpool Lime Street station. Entry to the museums is free, except for the Beatles Story.***

© SAKhan Photography / Getty Images

395

Savour every sandy second on Anse Marron

SEYCHELLES // You may have been hiking, bouldering, caving, bushwhacking and wading with your bag over your head for almost two hours to get here, but it's not the sunscreen and sweat causing you to rub your eyes – it's disbelief. In a nation famous for its princely beaches, Anse Marron is surely king. Brilliantly white sands meet crystal-clear pools, protected from the surging surf by Dalí-esque sculpted granite boulders. Schools of silver, near-translucent tropical fish dart in the shallows, so bring your snorkelling kit – though with scenery this stunning, it takes effort to look away and plunge your head underwater. The difficulty in accessing Anse Marron by foot only ensures that there are never more than a handful of people on the beach.

SEE IT! ***Anse Marron is at the southern tip of La Digue; a guide is necessary to get here.***

© 35007 / Getty Images

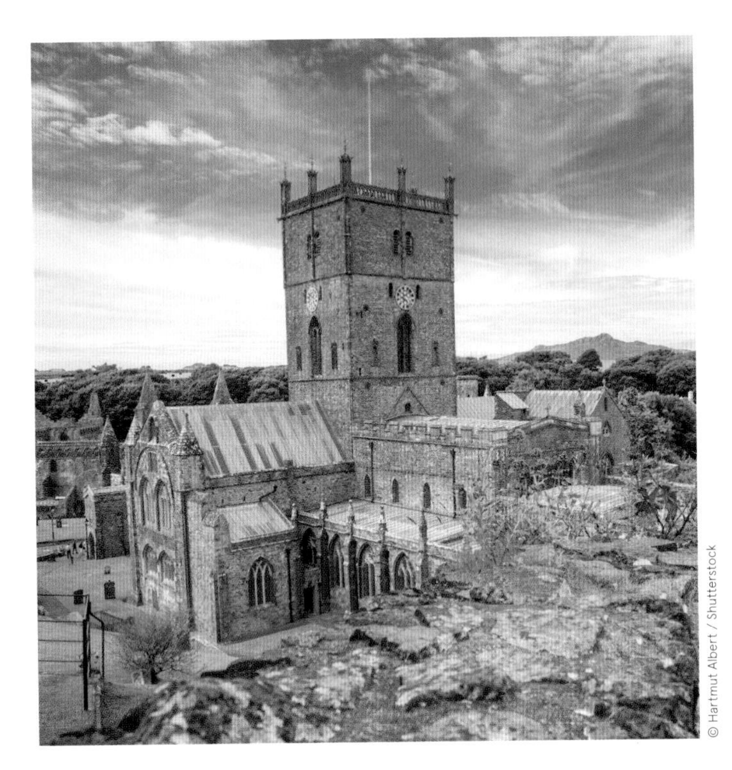

© Hartmut Albert / Shutterstock

396

Make the picture-perfect pilgrimage to St Davids Cathedral

WALES // Raised in purple Welsh sandstone, perched on a lovely limb of Pembrokeshire where Wales looks across the Celtic Sea to Ireland, St Davids Cathedral is immense, its scale way out of proportion to the tiny town surrounding it (the UK's smallest city). This is thanks to the miraculous performances of the local man who became St David – his bones are now behind the High Altar. It became one of Europe's most important Christian sites by the 12th century, prompting Pope Calixtus II to decree that two pilgrimages to St Davids were worth one to Rome.

SEE IT! *The cathedral, in the town of St Davids (Tyddewi), is a 2½-hour, 178km (110-mile) drive from Cardiff.*

© Tatiana Popova / Shutterstock

397

Experience life on the edge at Pulpit Rock

NORWAY // The sight of people perched right on the edge of Preikestolen (604m/1982ft), or Pulpit Rock, is one of Norway's signature images (there are no safety barriers, so watch your step). Hikers gasp at this finger of granite rock with its ethereal light and views extending deep into the inky blue waters of Lysefjord. Reached by a two-hour trudge, it's hands-down one of the most sensational lookouts on Earth, whether seen in the swirling mist or near-blinding Nordic sunshine.

SEE IT! ***Stavanger has domestic and international flights. Day trips to Pulpit Rock are possible with public transport.***

398

Discover the story of the unsinkable ship at Titanic Belfast

NORTHERN IRELAND // This museum stands on the site of the shipyard where the *Titanic* was built, but that's just the tip of the iceberg: the complex is a multimedia extravaganza that immerses you in marvellous minutiae covering every aspect of the ship's history via rides, talking holograms, rare film footage, poignant passenger accounts, computer 'fly-throughs' – even odours and heightened temperatures. It adds up to a multisensory experience that feels almost like sneaking aboard. The museum's gleaming, angular edifice occupies the head of the slipway from which the liner was launched. But although 'start' always leads to 'sink' in *Titanic* talk, the museum doesn't feel macabre, celebrating instead the achievements of ahead-of-their-time shipbuilders. As locals say, '*Titanic* was built by Irishmen and sunk by an Englishman!'

SEE IT! ***The museum is the centrepiece of the Titanic Quarter, the former shipyard area, an easy stroll from Belfast city centre.***

© VanderWolf Images / Shutterstock

399

Find humbling stories of survival at the War Childhood Museum

BOSNIA & HERCEGOVINA // A burned book, a handmade doll, humanitarian-aid food cans, ballet shoes... All exhibits in this deeply moving Sarajevo museum are personal belongings from survivors of the siege the city endured during the devastating 1990s civil war. Each token of a harrowing childhood experience is accompanied by an emotional reminiscence written by the person who donated it. The museum's concept was born out of a 2013 book edited by Jasminko Halilović, who asked the siege survivors to encapsulate their own memories of growing up during the Bosnian war. The result is an eclectic collection that tells a powerful story of suffering and hope, from the perspective of those most vulnerable and least responsible for the tragedy – the children of Sarajevo.

SEE IT! ***The museum is located in Sarajevo's historic centre; admission is free on the last Thursday of the month.***

© Oleksandr Rupeta / Alamy Stock Photo

400–500

© Sean Pavone / Shutterstock

© kaband / Shutterstock

400

Worship spectacular paintings in the Museo del Prado

SPAIN // One of the heavyweight galleries of European art and the definitive word on groundbreaking painters of Spanish origin, the Prado is a Madrid institution that can stand without blushing alongside the Louvre, the Uffizi and other illustrious collections of old masters. Housed in a noble neoclassical building that first opened in 1819, the museum's beautifully curated galleries allow unparalleled opportunities to contemplate the genius of Velázquez, Zurbarán, Murillo and – highlight of highlights – Francisco de Goya, whose fascinating oeuvre oscillates between reclining nudes (La Maja Desnuda) and violent executions (Third of May, 1808). Once you're done surveying the Spanish canvases, there's plenty more artistry to admire – Tintoretto, Rubens, Caravaggio, Rembrandt – the galleries read like a giant coffee-table guide to golden-age art.

SEE IT! *The Prado is in the centre of Madrid, 15 minutes' walk from Atocha train station. Book tickets online to avoid queues.*

401

Live the high life on Lake Geneva's stunning shores

SWITZERLAND // Western Europe's biggest lake is a beauty, backdropped by the French and Swiss Alps and flanked by alarmingly steep terraced vineyards. Turreted chateaux, luxe manor houses and ritzy hotels sprinkle its shores, all offering a slice of *la grande vie* (the high life).

While culture-rammed Geneva and upbeat Lausanne make terrific springboards for exploring, it's the little places beading the shores of Lac Léman (as it's known locally) that will really tug at your heartstrings. Top billing goes to the World Heritage–listed wine region of Lavaux, whose unique serried vineyards corduroy the hillsides. But Montreux, home to a legendary jazz festival, Vevey, with its sublime views, food and wine, and medieval Château de Chillon, should be way up on your wish list, too.

SEE IT! *Lake Geneva is in Switzerland's southwest. A Regional Pass will get you free and discounted travel on buses and trains.*

© Alexander Howard / Lonely Planet

© Lottie Davies / Lonely Planet

402

Meander past dramatic seas and cliffs on the Cabot Trail

CANADA // The Cabot Trail is a looping, diving, dipping rollercoaster of a road that snakes for 297km (185 miles) around the northern tip of Cape Breton, Nova Scotia. Get ready for heart-stopping sea views at every turn, breaching whales just offshore, moose nibbling at the roadside and plenty of places to stop and hike. Be sure to bring your dancing shoes – Celtic and Acadian communities dot the area, and their foot-stompin', crazy-fiddlin' music vibrates through local pubs. Meanwhile, artists' workshops are scattered along the trail's southeastern flank: there are pottery, glass and pewter workers, painters and sculptors, and you can also discover living remnants of Mi'kmaw and Acadian culture. Despite the winter snows, the trail is open year round.

SEE IT! *Most visitors drive west to east, starting in Chéticamp and ending in Ingonish, but there's less traffic and better views east to west.*

403

Go Med to the max on Croatia's fragrant Vis Island

CROATIA // It's the postcard image of a Mediterranean island: stark sea cliffs over azure waters, bobbing fishing boats, whitewashed houses with terracotta roofs, alleys with carved stone steps. Yet unlike some of its Greek neighbours, Vis remains pleasingly under-the-radar. Pebbly beaches such as Stiniva are reachable only by boat or by hairy scramble along a steep track. The hills, fragrant with rosemary and sage, frame views of ancient monasteries and fortresses. Cafes serve fresh seafood dishes enlivened with local olives and citrus; they pair nicely with white wines made from the local vugava grape. Off the southwestern tip, the island of Biševo is famed for the Blue Cave, whose waters glimmer an other-worldly blue when the sun hits in the late morning.

SEE IT! *The ferry from Split takes about 2½ hours; a catamaran is quicker. On the island, there's a bus between Komiža and Vis Town.*

404

Stand on Mt Kilimanjaro's summit, the snow-capped roof of Africa

TANZANIA // Stumbling across a permanent glacier when you're three degrees south of the equator is surprising, but climbing Kilimanjaro is an all-round astonishing experience. From the farmlands and lush rainforest of the mountain's lower slopes to alpine meadows and a barren lunar landscape on the push to the summit, the trek passes through a variety of microclimates. Once near the top, a highlight is beholding the plains far below, glowing in the sunrise. At 5896m (19,344ft), 'Kili' is Earth's highest free-standing mountain, and the elevation scuppers many attempts. Heed the words of Kilimanjaro's expert guides – *pole, pole* (slowly, slowly) – to maximise your chances.

SEE IT! *Moshi, Arusha and Machame are the gateways. Choose your trekking company and route carefully. Five-day treks are possible, but it's best to plan on six to eight nights to aid acclimatisation.*

405

Pay respect to a sacred relic at the Temple of the Tooth

SRI LANKA // The crowning jewel of Kandy town, the golden-roofed Temple of the Sacred Tooth draws pilgrims from across the world to visit Sri Lanka's most important Buddhist relic – one of the Buddha's canine teeth. The veneration provoked by this curio is profound and deeply sincere and captivating to witness. Don't expect to actually see the object of devotion: the tooth sits on a solid gold lotus flower, housed in a casket within a two-storey shrine and framed on either side by enormous elephant tusks. The shrine is open to visit during *puja* (prayers), which take place three times a day against a stirring beat of pounding drums and chanting.

SEE IT! *Kandy is 3½ hours from Colombo by bus. When visiting the temple, wear clothes that cover your legs and shoulders, and remove your shoes.*

© thinair28 / Getty Images

© Matt Munro / Lonely Planet

Top: horses graze on the shores of Son-Köl, a remote, unspoilt alpine lake in Kyrgyzstan. Bottom: Tending to the herd, Son-Köl

406

Sleep in a yurt on the shores of Son-Köl

KYRGYZSTAN // Decorated by a forest of mountain peaks and fringed by nomads' yurts, delightfully remote Son-Köl is one of Kyrgyzstan's most beautiful natural spots. And that's saying something in a country where unspoilt lakes, mountains and pastureland are the norm. Its sister lake, Issyk-Köl, is much bigger, much easier to reach and much more popular. But that's exactly what makes Son-Köl special. Situated at 3016m (9895ft), it's a summery dream when nomad families put up their camps for the season on the grassy *jailoo* (meadows) that front the lakeshore. Many participate in Kyrgyzstan's well-established community tourism network, opening their yurts as accommodation. During the day, hikes and horse rides offer varied vistas of the lake's gleaming waters. At night, a canopy of stars blankets the light-pollution-free skies.

SEE IT! ***The main cluster of yurtstays is at Batai-Aral on the north shore. The Community-Based Tourism office in Naryn can arrange transport and stays on the south shore.***

Opposite, from top: Visitors examine Georges Seurat's *La Grand Jatte* in Chicago; Elmina is one of Ghana's infamous slave forts

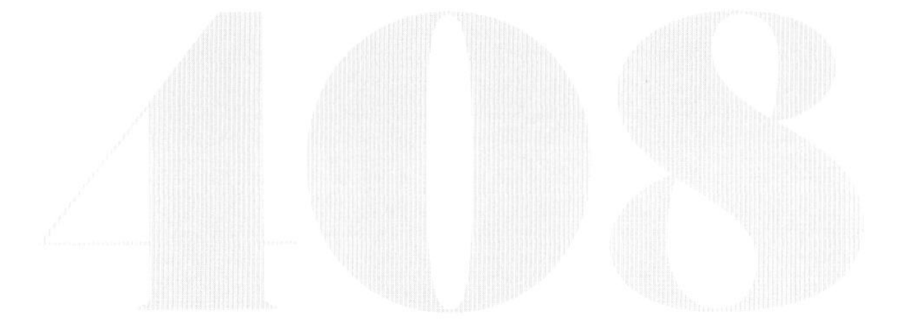

407

Admire superstar paintings in the Art Institute of Chicago

USA // Chicago's art museum has the kind of celebrity-heavy collection that draws gasps from patrons. Wander the endless marble corridors and Grant Wood's *American Gothic* appears (his sister and dentist were the models). Around the corner hangs Edward Hopper's lonely *Nighthawks*. Further on, Georges Seurat's big, dotted *Sunday Afternoon on the Island of La Grande Jatte* tricks your eyes. Then come Monet's *Stacks of Wheat*, Van Gogh's *The Bedroom*, Picasso's *The Old Guitarist* – the big-name roster goes on. The collection of impressionist and postimpressionist paintings is second only to those in France, and the number of surrealist works is mighty. Scads of spooky images by Salvador Dalí and René Magritte fill the Modern Wing, ready to haunt your dreams. But wait: we haven't even reached the odd bits and bobs, like the galleries stuffed with Japanese prints, Grecian urns and suits of armour. The basement holds rooms of miniatures – haven't you always wanted to see a tiny French boudoir circa 1740? – and 800 bejewelled glass paperweights. Sculptures and architectural relics stud the outdoor gardens.

SEE IT! ***The Art Institute sits in downtown Chicago. Public transport trains stop a block away. Time-sucking queues are rarely an issue.***

408

Contemplate sobering slave-trade history at Ghana's slave forts

GHANA // Ghana's slave forts induce a strong emotional response in all who visit. After all, they are a grim reminder of the transatlantic slave trade: countless souls arrived at these castles in chains, the majority having been sold by other West Africans to European traders, with the remainder captured directly. And if they survived the horrors of the long ocean journey, they had an unthinkably hard life of slavery ahead in the New World. On the other hand, the castles and the surrounding coastline have an eerie, undeniable beauty: with hulking whitewashed battlements watching over palm-shaded beaches where colourful fishing boats bob about. Spend time in Elmina and Cape Coast, where two of the largest slave trade forts still dominate the seashore, to make sense of it all: wandering beneath the ramparts, feasting on fried plantain on the beaches – all while pondering the fate of those who disappeared over the horizon and would never see these beautiful shores again.

SEE IT! ***Cape Coast and Elmina are the most popular castles. Cape Coast is two hours from Accra by bus; Elmina is a taxi ride from here.***

407

408

410

Feel the happy magic at Walt Disney World

USA // It's dubbed 'The Happiest Place on Earth' for a reason. Ride the Space Mountain coaster, glide through Epcot's Spaceship Earth, watch the fireworks over Cindarella's Castle, and we dare you not to grin. The Magic Kingdom is not just about American capitalism though; it's about a shared global culture. It's about Pakistani women wearing mouse ears atop their hijabs, and Chinese families picnicking on rice and fish at Tom Sawyer Island. It's an utterly immersive fantasy world, and you can't help but give in to the enchantment.

SEE IT! ***Orlando, Florida is the gateway to Disney. The park is crowded, but it's less so in May, and September to early November.***

409

Gaze at the towering bronze Tian Tan Buddha

CHINA // When flying into Hong Kong, take a peek out of the window and you may spot this giant Buddha perched atop Lantau Island. There are larger Buddha statues elsewhere in the world, but none that are seated, outdoors or made of bronze like this one, weighing 250 tonnes and towering over 34m (112ft) high. Slog your way heavenwards up 268 steps to his bird's-eye throne and you will appreciate Lantau as Buddha sees it: gloriously green and pretty.

SEE IT! ***You can reach Lantau by bus, metro or ferry from Hong Kong Island. Approach the statue via the scenic bayside cable car.***

411

See surrealism on steroids at Teatre-Museu Dalí

SPAIN // Take the most vivid dream you've ever had and multiply it by 10,000. Spiralling from the madcap imagination of the 20th century's showiest surrealist, this museum was designed by Dalí himself in 1968. The exhibits reflect the personality of their twirly-moustached creator. Giant eggs adorn the rooftop, baguette-topped figurines guard the balconies; mirrored flamingos, a supersized pair of lips and Dalí's own crypt greet you inside. Incredible is an understatement.

SEE IT! ***The museum is in the town of Figueres, 1¼ hours from Barcelona by train.***

© kovop58 / Shutterstock

Top: sunset in the High Tatras. Bottom: The jagged peaks of the High Tatras from Poland's Pieniny Mountains

412

Spot bears and scale peaks in the High Tatras mountains

SLOVAKIA // Slovakia's loftiest entry on this list is the fittingly named High Tatras: a slew of jagged peaks of which 25 surpass 2500m (8200ft). Slovakia's and the entire Carpathian range's highest point is here at Gerlachovský štít (2655m/8710ft) and this is one of Europe's few terrains with Alpine characteristics outside of the Alps. Cue some magical wildlife found in few other European locales, including three of Europe's Big Five: the brown bear, wolf and Eurasian lynx. Hiking is high on priority lists here, with the long-distance Tatranská Magistrala trail traversing the range. But best of all might just be bedding down in one of many high-altitude wilderness lodges and waking up to other-worldly summits sweeping around you like the continuation of a dream.

SEE IT! *Poprad, 330km (205 miles) south-west of Bratislava, is the nearest big city. A mountain railway climbs into the High Tatras.*

© Carlos Amarillo / Shutterstock

© Alvaro Trabazo Rivas / Shutterstock

© alessandro_pinto / Getty Images

413

Test your nerve on the cliff-hugging Caminito del Rey

SPAIN // The Caminito del Rey was never meant to be a tourist attraction; it was built in 1905 to allow workers to access the hydroelectric power plants that were wedged into this impossibly steep gorge. It was crowned with its current name, the King's Path, after King Alfonso XIII walked it to inaugurate the dam in 1921. By the 1990s, the path had fallen into severe disrepair – entire sections of concrete and handrails had dropped off the rockface – and was labelled the most dangerous hike in the world after several daredevils attempted to traverse it and not all survived. After an extensive overhaul, a new path made of wooden slats opened in 2015. The views of the vertiginous gorge and the walkway clinging to the cliff as it snakes through the canyon are out of this world.

SEE IT! *Trains connect Málaga to El Chorro (40 minutes), then it's a walk to the start. Pre-book online and allow at least half a day.*

414

Dip into the Unesco-listed splendour of Quito Old Town

ECUADOR // That Quito Old Town, the Ecuadorian capital's crown jewel, was made Unesco's first-ever cultural World Heritage site speaks volumes about this lavish spread of restored colonial architecture sheltering below vertiginous volcanic peaks. Stippled by sumptuous churches, punctuated by pretty plazas and blessed with innumerable spots to soak in the views, the historic core of this high-altitude Andean city is a place to enjoy on foot, strolling cobbled thoroughfares, staring up at spires and swinging by magnificent museums. Plaza Grande is a good spot to begin. Its cathedral is graced by paintings from famous Quito School artists and its grand Palacio de Gobierno, seat of the Ecuadorian presidency, sports a mosaic by Oswaldo Guayasamin depicting Francisco de Orellana's original descent of the Amazon.

SEE IT! *Quito's airport has flights from destinations across North, Central and South America and a handful of European cities too.*

415

Step into a samurai stronghold at Himeji Castle

JAPAN // The graceful contours and lustrous white facade make multi-tiered Himeji-jō look more like a wedding cake than a military fortress for samurai and shoguns. Japan's most magnificent castle is the best example of only a dozen originals in the country – many others are concrete reconstructions, having been destroyed by natural disasters or war. More than 400 years old, the castle complex is huge; with the soaring five-storey main keep, there are more than 80 buildings, accessed through 21 gates (of an original 84), all surrounded by moats and defensive walls. While walking through, you'll get a lesson in medieval defensive strategies, like the *ishiotoshi* – narrow openings that allowed defenders to pour boiling water or oil onto anyone trying to scale the walls. Now you can go straight in, but check the castle website's congestion forecast calendar for the best days.

SEE IT! ***The small city of Himeji is under an hour from Kyoto by train.***

Doina Ciobanu's Top Five Places

Model, entrepreneur and 'No More Plastic' ambassador Doina Ciobanu harnesses her digital platforms to push for environmental causes, including sustainability in the fashion industry.

01

HYDRA ISLAND, GREECE – The lack of cars and the quiet streets where you can walk barefoot make you feel you've travelled back in time. There's amazing food, from ice cream to pizza and traditional Greek tavernas. Make sure you have a souvlaki at Kremmidi.

02

POINTE MILLERS, ANSE GEORGETTE BEACH, SEYCHELLES – Never have I seen dolphins swim freely so close to the shore, along with the most colourful fish. And there's no development. It is a pure piece of wild heaven.

03

THE ROOF OF THE GRAND BAZAAR, ISTANBUL – One of my favourite places in the world. The quietness above such a buzzing city is unique. Especially during prayer time, when the air fills with echoes from hundreds of mosques.

04

KYOTO – I know it's a busy city but it's my favourite. With 1600 temples, it is a place to explore by bicycle and stop at all the little gardens and forests on the way. I'd recommend visiting in spring during Sakura (cherry blossom season).

05

HANG-GLIDING IN RIO – The views of Rio from above are unforgettable. I went up with one of Rio's most experienced gliders, Carlos Eduardo Renha da Rocha. Honestly, it was perfect, and not so scary when you have a good instructor.

© D'July / Shutterstock

© Sandro Helmann / Getty Images

416

Explore medieval Dravidian skyscrapers at Virupaksha Temple

INDIA // Hampi's Virupaksha Temple and bewitching ancient city ruins can be an unreal place to linger, set amid a landscape marbled with giant boulders, and devoid of villagers since a 2012 court ruling. The intricately carved, 15th-century *gopuram* (gateway tower), which is almost 50m (165ft) high, is a superb example of southern Indian temple architecture and the symbol of a lost city that once housed half a million residents as the capital of the last great Hindu kingdom of Vijayanagar. With 3700 monuments to explore in Hampi, from elephant stables and royal bathtubs to sacred temples and opulent residences, it's worth days of your time. If you tire of the ruins, Hampi is India's bouldering capital and a great place to learn the sport.

SEE IT! ***Bangalore and Goa have international airports; rail links to nearby Hospet abound. Stay in Virupapur Gaddi, across the river.***

417

Drift through clear waters and dripping karst in Bonito

BRAZIL // Authorities take sustainability seriously in this watery Eden, where you can swim beneath waterfalls, dive in caves with massive underwater stalagmites, and float down rivers clear as glass beneath a pinkening tropical sky. That means popular attractions have strict visitor limits, so book ahead. Found in the landlocked state of Mato Grosso do Sul, Bonito owes its landscape to its Swiss cheese-like karst geology, where rock eaten away by eons of rainfall turns into caves and sinkholes. There are few places in the world with rivers clear enough to snorkel in. Lay on your stomach and let the current move you as tropical fish flit past your mask, while the chatter of capuchin monkeys in the trees above is dulled by the water's burble.

SEE IT! ***The bus from Campo Grande takes about five hours; there are limited flights. All sights must be booked ahead with tour operators.***

418

Understand courage at the Mississippi Civil Rights Museum

USA // This new museum in Jackson, the state capital, presents one of the truest looks at the struggle for racial equality you'll find anywhere. Mississippi was on the front line of the national Civil Rights movement, and exhibits delve into headline-capturing events that happened there such as the murder of Emmett Till and the assassination of Medgar Evers. The low-lit galleries brim with films, recordings and interactive displays that put your senses on high alert, whether it's graphic photos of lynchings that hit you with a gut punch, or a recreated jail cell that immerses you in the brutal conditions activists faced. The museum doesn't sugarcoat its story of segregation, protest, progress made and challenges that remain. And therein lies its power.

SEE IT! *The museum is in downtown Jackson and open Tuesday to Sunday. It shares a building with the Museum of Mississippi History.*

© Tom Beck

419

Sail to the kingdom of the kiwi on Stewart Island

NEW ZEALAND // New Zealand's third island is way down south, separated from the mainland by choppy Foveaux Strait. It's a classic end-of-the-earth environment, where nature reigns and a handful of hardy humans fit in around it.

Rakiura National Park makes up 85% of the land, which ranges from long beaches and sheltered inlets to lush forest carpeted in ferns and moss. It's a largely pristine and remote world, and an internationally renowned bird haven. This is the kingdom of the kiwi, New Zealand's endangered national icon. At around 20,000 birds they outnumber locals by over 50 to one and are so relaxed they can occasionally be glimpsed before sundown, despite being nocturnal and notoriously shy.

SEE IT! *An hour's ferry ride from Invercargill, Stewart Island is best visited on a grand tour of Fiordland and the Southern Lakes.*

© R. Vickers / Shutterstock

© Sean Pavone / Shutterstock

Griffith Observatory looks out to the rest of the cosmos, as well as across to the bright lights of Los Angeles

420

Stargaze in more ways than one at the Griffith Observatory

USA // Plonked down on the slopes of Mt Hollywood, this gorgeous art deco institution was a movie location waiting to happen. Indeed, it features in dozens of Hollywood films, from *Rebel Without a Cause* to *Transformers* and *The Terminator*. The sci-fi connection is actually tied to science fact – the observatory has been scanning the stars for evidence of life beyond earth since before WWII. Today the world's most advanced star projector is here, along with historic telescopes and high-tech shows that peer into the cosmos. Don't forget to look down as well as up. The venue's location unfurls sweeping views that take in Los Angeles' skyscrapers and mansions, endless avenues, rugged hills and the legendary Hollywood sign.

SEE IT! *Griffith Observatory sits on the southern slopes of Griffith Park; hike here from Los Feliz, or ride the bus from Vermont/Sunset Metro Red Line station. Crowds amass on weekends.*

421

Encounter dragons on Komodo

INDONESIA // From the rugged beauty of its volcanic islands to its magnificent coral reefs, Komodo National Park is one of Indonesia's most dramatic landscapes. But this magical place, nestled between Sumbawa and Flores in the centre of the Indonesian archipelago, is best known for its Komodo dragons. The endangered lizards can reach up to 3m (10ft) in length, weigh up to 100kg (220lb) and take down a deer with their venomous bites and dagger-sharp claws. About 5700 of them inhabit the region and should be observed from a safe distance! Throw in some seriously world-class diving, and you've a traveller's dream.

SEE IT! ***Day trips can be booked from Flores, though it's worth staying in the park overnight. There are hefty fees for park entry.***

© Danaan / Shutterstock

Hook onto Norway's seafaring heritage in pretty-as-a-picture Bryggen

NORWAY // Shaped by seven fjords and seven hills, Bergen is the Norway of your wildest dreams. The rust-red wooden wharf houses of its old town, Bryggen, create a warm glow along the harbour front. This was an important trading hub in the 14th to 16th centuries, and the atmosphere of an intimate waterfront community remains; losing yourself among the gabled buildings and alleyways is one of Norway's greatest pleasures.

SEE IT! ***Bergen lies amid Norway's southwestern fjords. Visit during shoulder seasons to avoid the cruise ship and tour bus rush.***

423

Wander through the green grounds of Dublin's Trinity College

IRELAND // The gorgeous Georgian buildings, greens and cobbled squares that sprawl around prestigious Trinity College, alma mater of Jonathan Swift, Oscar Wilde and Samuel Beckett, combine to create one of Dublin's loveliest areas for strolling through. Visit the Old Library to see a wealth of treasures, including the world-famous *Book of Kells*, a stunning illustrated manuscript produced in the 8th or 9th century.

SEE IT! ***It's free to wander around the grounds. The Old Library gets busy during the summer months; book online to skip the queue.***

© Lottie Davies / Lonely Planet

Aït Ben Haddou began its life as a roadside inn a thousand years ago, and is now Unesco-protected

424

Tour Lawrence of Arabia's Moroccan mud fortress at Aït Ben Haddou

MOROCCO // Art imitates life at Aït Ben Haddou, rising up from the southern foothills of Morocco's Atlas Mountains like some wildly imagined fortress from *The Arabian Nights*. Movie buffs may recognise it from *Lawrence of Arabia*, *Jesus of Nazareth* (much of it was rebuilt for the production) and *Gladiator*, all of which used this red mud-brick *ksar* (fortified village) as their backdrop. While film-maker interest saved it from ruin, Unesco protection has helped preserve Aït Ben Haddou, which was born in the 11th century as an Almoravid caravanserai. Today you'll find a tangle of medieval lanes climbing to a summit from where a panorama of mountains, palm groves and stony desert stretches to the horizon.

SEE IT! ***Aït Ben Haddou is a short hop from the southern Atlas town of Ouarzazate, which has transport to the rest of the country.***

© Suzi Pratt / 500px

© Rolf E. Staerk / Shutterstock

425

Time-travel to the Mayan treasure of Xunantunich

BELIZE // Travel is more about the journey than the destination, and that's particularly true when arriving at Xunantunich, one of Belize's most important Mayan sites, via hand-cranked cable ferry. Upon being deposited on the western side of the Mopan River, visitors need only walk or drive 1.6km (1 mile) from the riverbank to reach these revered ruins, occupied as early as 1000 BC by as many as 10,000 villagers. Here are grand plazas, carved stone monuments and the breathtaking 40m (130ft) pyramid El Castillo: climb to its top, survey all the land and declare that everything the light touches is your kingdom. After all, you're standing on the ruling family's ancestral shrine, or at least the stucco frieze down the eastern side seems to suggest as much. Don't miss the visitor centre, which houses displays on the site's rich history.

SEE IT! *Support the local community by hiring a guide. They are highly knowledgeable and will enliven your experience.*

426

Explore the witch's hat-turreted walled city of Carcassonne

FRANCE // Carcassonne's concentric walled city (La Cité) is a fairy-tale vision of a medieval castle that has been a Unesco World Heritage site since 1997. Perched on a rocky hilltop in the Languedoc region and bristling with battlements and conical turrets, the fortified city looks like something out of a children's storybook. The mystical atmosphere is turned up a notch at night when the old city is illuminated and glowing.

In summer, La Cité's cobbled streets and squares spill over with souvenir shops, cafes and huge tourist crowds. To savour its charms, cross the moat to enter its main gate of Porte Narbonnaise at dawn, or linger at dusk when it belongs to its few inhabitants and visitors staying within the ramparts.

SEE IT! *Cruise along the Canal du Midi from Port de Plaisance in the Ville Basse (lower town) to appreciate the architecture from afar.*

Aldo Kane's Top Five Places

Scottish adventurer Aldo Kane is a world record holder and former Royal Marines commando sniper. He has operated and filmed in over 100 countries, been held at gunpoint, charged by a black rhino, abseiled into an active volcano, escaped Ebola, rowed the Atlantic and dived on Captain Kidd's ship.

01

THE HIDDEN VALLEY, GLEN COE, SCOTLAND – Steeped in history, this is one of my favourite low-level walks in Scotland. Take the footpath from the road up to the large glacial valley; most people can walk it in under an hour.

02

LOUTRO, CRETE – Loutro is nestled in a secluded bay where the Cretan mountains meet the sea. I always stay at the Hotel Porto Loutro and lunch at the Blue House. Loutro has everything for me: climbing, swimming, walking and relaxing after busy expeditions.

03

VIRUNGA NATIONAL PARK, DRC – One of the most beautiful areas on the planet. There are three things you must see: Lake Kivu, the Mountain Gorillas and the spellbinding lava lake of Mount Nyiragongo Volcano.

04

SURINAME – One of the most remote, interesting places I've been. Kabalebo eco-lodge is set so deep in the forest most of its wildlife – macaws, harpy eagles, jaguars and giant tapirs – has never seen humans before.

05

BARRA, OUTER HEBRIDES, SCOTLAND – A five-hour ferry from Oban, Barra's white sand beaches rival any Caribbean or Pacific island I have been to. More often than not, you'll have it completely to yourself. My favourite time to visit is January – the sea is wild.

427

See stars in the cracked Desierto de la Tatacoa

© Jess Kraft / Shutterstock

COLOMBIA // Tatacoa looks like it should have its picture above the word 'desert' in the encyclopedia. Deeply wrinkled gullies of cracked ochre earth. Mesas topped with crooked cacti. A pitiless sun. Rattlesnakes. And yet this is not technically a desert, but a dry tropical forest ringed with mountains. Come to this unique landscape to spot lizards, wildcats and dozens of bird species, hike or bike to sunset-viewing spots, and ride horses through the silent canyons. The absence of urban light pollution means it's one of the world's finest star-spotting locations. Visit one of two astronomical observatories to hear stargazing talks and use the telescopes. Don't miss the lesser-visited grey area of Tatacoa – it's like walking on the moon.

SEE IT! *The city of Neiva has buses and flights to Bogotá, Cali and Medellín. It's a short bus trip to the access town of Villavieja.*

© Veronika Kovalenko / Getty Images

© Jonathan Gregson / Lonely Planet

428

Descend into the hallowed halls of Kyevo-Pecherska Lavra

UKRAINE // Set in lush grounds beside the Dnipro River, the Lavra is the definitive Eastern Orthodox monastery, a cluster of shining golden domes and soaring arches graced with mosaics of the saints. Founded by the Greek St Antony and his followers (who now spend eternity in the catacombs they dug out below the lower church), this baroque beauty is a living vision of Kyivan Rus. The Lavra hides genuine treasure – a magnificent hoard of Scythian gold gathered from the tombs of the ancient horsemen of the steppes. The real surprise, though, is the basement, where the monastery's founders lie in elegant coffins, their bodies desiccated but preserved by the caves' cool and dry conditions. Mobs of pilgrims fingering rosaries and kissing the mummies and icons only add to the atmosphere.

SEE IT! *The Lavra is accessible by metro from downtown Kyiv; disembark at Dnipro or Arsenalna and walk south through the park.*

429

Wildlife-watch in cloud cuckoo land at Bosque Nuboso Monteverde

COSTA RICA // Even the name of this Costa Rican oasis – 'green mountain' cloud forest – evokes a mystical place, where life is abundant but concealed. Soaring trees are so draped in epiphytes that it's impossible to tell forest floor from canopy. The lushness is seemingly impenetrable, but punctuated by orchids and flame-coloured bromeliads. The silence is amplified by the drip, drip, dripping from the canopy – then shattered by the clang of the three-wattled bellbird. Monteverde sits atop the Continental Divide, providing a habitat for hundreds of species of birds and animals, such as the quetzal, spiny mouse and tapir. Swirling with mist and seething with life, Monteverde has an other-worldly aura. It's no wonder scientists and Quaker farmers alike have been moved to cooperate in order to protect and preserve it.

SEE IT! *The main entrance is 6km (4 miles) south of Santa Elena in Puntarenas. It rains least (but still lots) from December to April.*

© grantreig / Getty Images

Seemingly endless stretches of marshes, dunes and beaches stretch out to the Atlantic in wild Cape Cod National Seashore

430

Enjoy dunes, whales and oysters at Cape Cod National Seashore

USA // Thank President John F Kennedy for this unspoiled crook of land jutting out from southeastern Massachusetts. He preserved some 64km (40 miles) of his home-state coastline, so it remains an endless vista of sand and sky where beaches, dunes, salt marshes, nature trails and forests rule the wild Atlantic landscape. Beachcombers wade among sandbars, surfers ride the waves, and hikers and bikers hop on trails edged by saltspray roses, scrub pine and cranberry bogs. Boat tours go out to spot humpback whales, as the area just offshore is the behemoths' summer feeding ground. Speaking of food: more clam shacks than you can shake a napkin at are cooking along the shore, prime for creamy chowder, fried clams and sweet, briny oysters fresh from local waters. Atmospheric towns enhance all the nature, such as Wellfleet (where those slurpable oysters come from) and flamboyant, gallery-laden Provincetown. Most of the area slumbers from mid-fall to late spring, then explodes with visitors from roughly June to September.

SEE IT! *The seashore's entrance at Eastham is a 150km (93-mile) drive from Boston; traffic can be brutal. Fast ferries run from Boston to Provincetown in 90 minutes.*

© WaterFrame / Alamy Stock Photo

431

Dive into a sheer shark abyss at Hammerhead Point

MALDIVES // Just to be clear: Hammerhead sharks rarely bite humans and have never killed anyone. So this bucket list dive is about awe, not adrenaline. And what awe! Dawn dives at Madivaru Corner, aka Hammerhead Point, mean hovering in the silent, half-dark waters waiting for a car-length hammerhead to emerge from the gloom and scrutinise you with its strange side eyes. From January to March, ocean currents push nutrients up from the deep, feeding the fish that in turn become hammerhead chow. The oceans are also clearest at this time of year, meaning dizzying visibility into the 200m (655ft) abyss below. Even if you miss the hammerheads, you'll still be joined by barracudas, grey reef sharks, dogtooth tuna and more along the outer reef.

SEE IT! ***Hammerhead Point is in the Maldives' Rasdhoo Atoll, reachable by seaplane or ferry from Male or as part of a liveaboard trip.***

© Mike Mareen / Shutterstock

432

Behold a king's whimsical creations at Potsdam's rococo Schloss Sanssouci

GERMANY // Just imagine the kind of summer palace you could build with unlimited dosh, a fervent imagination and the best architects of the age. It might look something like Schloss Sanssouci. Terraced vineyards stagger up to this regally rococo, butter-yellow delight of a palace. Designed by Georg Wenzeslaus von Knobelsdorff in 1747, it was the much-loved retreat of Frederick the Great (1712–86), who craved somewhere he could escape *sans souci* (without cares). Beyond the frescoed and gilded interiors of the Neue Kammern (New Chambers), allow time for the old masters that hang in the Bildergalerie, as well as the fountain-splashed park. The king's French-style pleasure garden hides a Chinese tea pavilion and shell-encrusted Neptune grotto.

SEE IT! ***Potsdam is a 25-minute train ride southwest of Berlin; some trains continue to Sanssouci. Book online – admission is by timed ticket.***

433

Float through *Avatar*'s mountains in Zhangjiajie

CHINA // The quartz spires and soaring craggy peaks of Zhangjiajie National Forest Park are said to have been the inspiration for Pandora in the movie *Avatar* and it's not difficult to imagine why. The mountains here rise into strange columns of rock dusted by greenery, their peaks shrouded in clouds. This is the stuff of classical Chinese paintings – if a dragon soared by, would you even look twice? The 4810-hectare (11,900-acre) national park is a Unesco-listed global geopark that forms part of the giant Wulingyuan World Heritage Site. Within are hundreds of peaks, caves, trails and valleys to explore.

The mythical pillar-peaks here may feel like something outside of time and space, but Zhangjiajie is also full of modern superlatives. The world's tallest outdoor lift – Bailong ('Hundred Dragons') Elevator – zips sightseers up and down a sheer cliff face. The planet's longest and highest pedestrian glass bridge offers views of Zhangjiajie Grand Canyon from on high. And the world's second-longest cable car floats up a staggering 7.5km (5 miles) from downtown Zhangjiajie to the top of Tianmen Mountain.

SEE IT! ***The park is a four-hour bus journey from Hunan's capital, Changsha. You can stay in Wulingyuan or Zhangjiajie.***

The sandstone pillars of Zhangjiajie National Forest Park are the stuff of classical paintings and are said to have inspired *Avatar*

434

Gaze across the plains from clifftop Prasat Preah Vihear

CAMBODIA // Dating from the 11th century and dedicated to the Hindu deity Shiva, Prasat Preah Vihear has the history, monumental architecture and exquisite carvings that are highlights of other temples of the mighty Khmer Empire, but it is the dramatic location that sets it apart. The temple sits high on an escarpment in the Dangrek Mountains, its foundations reaching right to the cliff edge. From here you can survey the plains of northern Cambodia for miles around. Bring a picnic.

SEE IT! *It's 3 hours by car or tour from Siem Reap. Tensions between Cambodia and Thailand can flare – check the situation before you go.*

435

Witness a feeding frenzy in Shark Ray Alley

BELIZE // Now-protected Shark Ray Alley began as a place for local fishers to clean their catches, tossing away bits of fish that were delicious to nurse sharks and stingrays. With Pavlovian regularity, the hungry (but harmless) sea creatures returned, eventually gliding in at the mere sound of a boat, and soon those boats belonged to snorkel captains. These days, snorkellers can observe the frenzy (no touching, of course) in just 3m (10ft) of impossibly clear, warm water. Visitors will also be wowed by the surrounding shark-infested coral reefs.

SEE IT! *You'll find plenty of responsible operators in Ambergris Caye. Sunscreen can damage the ecosystem; cover skin up instead.*

436

Watch the guanacos roam in Patagonia National Park

CHILE // South America's largest camelids, doe-eyed guanacos roam the steppe in herds – when not lounging tamely by the visitor centre. It's hard to fathom that this 2800-sq-km (1080-sq-mile) jewel was a dusty working ranch populated by sheep as recently as 2004. The vision of conservationist Kristine Tompkins, it has world-class trails and wildlife watching in ecosystems ranging from Patagonian steppe to alpine peaks, forest and lakes. Tompkins Conservation is rewilding rheas and Andean condors while monitoring the endangered Andean deer and pumas.

SEE IT! *It's utterly remote. The closest airport, Balmaceda, is six hours north. Buses are infrequent, and travelling by 4WD is best.*

437

Delve into history at St Fagans National History Museum

WALES // For a lesson in Welshness, St Fagans provides a microcosm of life in Wales like no other. Boring history lesson this is not. In this living museum of more than 40 original buildings, you can sneak inside still-smoke-scented 16th-century farmhouses, time-travel through miners' cottages, marvel at an ancient church moved here stone by stone and behold the reconstructed 12th-century court of Welsh titan Llywelyn the Great. The display is anchored by a medieval castle worthy of its eclectic dominion.

SEE IT! *St Fagans is 7.5km (5 miles) west of central Cardiff (25 minutes by bus).*

© Mark Read / Lonely Planet

Cape fur seals lounge in the sun in Namibia's Cape Cross National Reserve – the largest breeding colony for the seals on the planet

438

Watch seals frolicking in cold Atlantic waters at Cape Cross Seal Reserve

NAMIBIA // As you approach the headland housing the world's largest breeding colony of Cape fur seals, you'll first hear the barking, and then notice the odour. Very rotund from an abundant diet of fish, and very uninhibited, the seals lounge all over the beach and rocks, some individually, others piled on top of each other, and they fill the offshore waters. In peak breeding season from November to December, their numbers can reach 200,000. Don't attempt to touch them. Apart from not being sound practice environmentally, seals can be temperamental, especially when on land, and have been known to nip humans who venture too close. Near the main seal area is a replica of the stone cross erected in 1485 by Portuguese explorer Diogo Cão, which gives the headland its name.

SEE IT! ***Visit as a day trip from Swakopmund – the main gateway town – or, better, as part of a longer Skeleton Coast itinerary.***

439

Learn from the past at Nagasaki Atomic Bomb Museum

JAPAN // As one of the few ports open to foreign trade during two centuries of Japan's self-imposed isolation, the city of Nagasaki has long been historically significant. But for most it is known for 9 August 1945, when it became the second city in the world to be devastated by an atomic bomb. Its sombre Atomic Bomb Museum recounts the destruction through salvaged remains and photos, but perhaps most moving are the firsthand accounts from survivors, who tell stories of both great terror and great heroism.

SEE IT! ***The museum can be reached by streetcar from Nagasaki Station; the hypocentre and Peace Park are nearby.***

440

Find unearthed Aztec treasure at Templo Mayor

MEXICO // To the Aztecs it was centre of the universe, a 60m-high (200ft) temple dedicated to the gods that towered above their sophisticated capital, Tenochtitlán, until the Spanish destroyed it in 1521. But some legends refuse to die. Unearthed by electricity workers in 1978, the temple's ruins have so far yielded over 7000 artefacts and helped piece together the story of the hulking structure that once stood here. Precious booty is laid out in an eight-room on-site museum that archives the history of Mesoamerica's greatest monument.

SEE IT! ***The ruins and museum are conveniently located in the northeast corner of Mexico City's central Zócalo.***

441

© Balate Dorin / Shutterstock

441

Gawp at Bucharest's Palace of Parliament

ROMANIA // This 1100-room colossus was designed to inspire fear and awe. From 1984, more than 100,000 workers toiled through the night to fashion its floors, using one million cubic metres (over 35 million cubic ft) of marble. Hung from the stuccoed ceilings are 2800 chandeliers. Imposing neoclassical details and Socialist Realist emblems are everywhere. Pleasingly, the overblown edifice now houses such engines of democracy as the Romanian Senate – a tonic to bitter memories of the legacy of Nicolae Ceaușescu, overthrown and executed in 1989.

SEE IT! ***Walk from Bucharest's metro station Piața Unirii for imposing views of the palace. Bring your passport for entry.***

442

Worshippers congregate around the Kaaba, a granite chamber in the Great Mosque of Mecca that is the holiest site in Islam

Have a spiritual experience in Mecca

SAUDI ARABIA // Mecca is off limits to non-Muslims. But to those who practise Islam, the desert city is the holiest of holy places. Its spiritual centre is the ornate white and green Al Masjid Al Haram – the Great Mosque of Mecca – the biggest mosque in the world. Spread across 356,800 sq metres (3.8 million sq ft), it can hold a million worshippers. It wraps around the Kaaba, the black- and gold-covered cube that is the holiest site in Islam, sometimes called the House of God. When Muslims around the world pray, they orient their prayer mats towards the Kaaba. Though the mosque is ancient, the skyscrapers throwing shadows over the Kaaba remind you that this is very much the 21st century.

During the hajj, Mecca hosts one of the world's largest human gatherings. All Muslims who are physically and financially able are expected to journey to Mecca at least once in their lifetimes; each year about 2.5 million answer the call. The city becomes a teeming sea of white-clad pilgrims, praying, drinking from the Zamzam Well, and sleeping everywhere, from luxury hotels to Japanese-style nap capsules.

SEE IT! *The nearest airport is in Jeddah, with trains and buses running to Mecca. Non-Muslims are forbidden to enter the city.*

© David Ionut / Shutterstock

443

Enter a fairy tale of Old York at York Minster

ENGLAND // York's ancient walls circle a city steeped in legend, and its mighty medieval minster is as storied as they come. Northern Europe's largest Gothic cathedral, it's home to half the medieval stained glass in England, and is the resting place of Viking kings and English nobles. Begin in the Undercroft museum, which leads you through 2000 years of local history, from the remains of Roman barracks to the foundations of the Norman building you're standing in. Next door, the treasury contains the Horn of Ulf, a thousand-year-old elephant tusk that was carved in Italy and used as a cup for beer or mead. Finish by climbing the cathedral's central tower for cracking views across Yorkshire.

SEE IT! ***York Minster is in the centre of the city. Admission tickets last a year – so you can take multiple visits.***

© halfofmoon /Shutterstock

444

Search out a pristine Pacific island reserve at Parque Nacional Coiba

PANAMA // Sometimes crime can pay, at least if you're a bird, fish or turtle. The Panamanian island of Coiba, 20km (12 miles) from the mainland, was once a notorious penal colony with little infrastructure and fewer visitors – a situation that left its rich waters almost completely untouched and its wildlife thriving.

It's now a World Heritage–listed national park, and visitors can hike through primary rainforest and snorkel and dive on the surrounding reefs, which host huge schools of fish. Permits are required to visit and infrastructure is still poor, so you need to be self-supporting here, with the help of a local tour operator.

SEE IT! ***The ranger's station that marks the entrance to Coiba is a one-hour boat ride from the coastal city of Santa Catalina.***

445

Kayak the wildlife-rich waters of Isla Espíritu Santo

© Miguel Zetter Lopez / Shutterstock

MEXICO // From a potential environmental meltdown springs hope. Lapped by the gentle waves of the Sea of Cortez in the state of Baja California Sur, uninhabited Espíritu Santo came close to being blemished by a casino resort in the 1990s. Fortunately, eco-warriors prevailed over real estate developers, meaning the 81-sq-km (31-sq-mile) island, now carefully protected by Unesco, has safeguarded its pristine collection of diamond-dust beaches, scimitar-shaped bays and diverse marine life guarded in what they call 'the world's largest aquarium'. Disembark on land and it's possible to follow ten different hiking trails in search of the threatened black jackrabbit, a mammal unique to the island. But, for the best activities, stay in the water, interspersing kayaking around rose-pink cliffs with snorkelling in cerulean seas awash with fish.

☛ SEE IT! ***Numerous companies offer day trips to Espíritu Santo out of the city of La Paz.***

Tima Deryan's Top Five Places

At only 26 years old, entrepreneur and climber Tima Deryan became the first Lebanese woman to summit Mount Everest. She has now scaled six of the Seven Summits.

TOUR DU MONT BLANC – If you can spare nine days for an adventure on foot, Tour Du Mont Blanc takes the top spot. The trail winds round Mont Blanc through France, Switzerland and Italy. It's well marked, and you can do it solo.

RAS AL KHAIMAH – I trained for my Everest climb on Jebel Jais in Ras Al Khaimah. It's the highest peak in the UAE and has beautiful views of desert, mountains and sea. It took my breath away. The Emirate is transforming into the adventure hub of the Middle East.

TALKEETNA GUEST HOUSES, ALASKA – One of the smallest and most unique towns I've visited. It's so vibrant and showcases the beautiful scenery of the highest peak in North America – Mt Denali. In a cosy guest house you get to experience true American culture.

LEBANON TANNOURINE – Tannourine is the biggest crag in Lebanon and has routes ranging from 5a to 8b, making it an ideal spot for beginners and professional climbers. It's a great escape from the city heat and pollution.

LAGUNA VERDE, CHILE – After coming down from the summit of Ojos Del Salado, I took a dip in the Laguna Verde salt lake. It's around 4000m up in the Chilean Andes. You can set a tent next to the lake and spend the night.

© Matt Munro / Lonely Planet

© Luciano Lejtman / Alamy Stock Photo

446

Bike and hike in Seyðisfjörður & Borgarfjörður Eystri

ICELAND // Fjord-lovers, prepare to worship this little pocket of East Iceland. Surrounded by mountains and waterfalls and nestled at the bottom of a steep valley, stunning Seyðisfjörður rewards all those who make the (sometimes challenging) trip into her secluded haven. Kayakers, mountain-bikers and hikers are in heaven here, hikers even more so at nearby Borgarfjörður Eystri, with its spectacular routes untouched by the masses, its sustainable-tourism focus and its waterside hot-pots for post-hiking relaxation. It's also home to an easily accessible islet covered in puffin burrows, a perfect place to spot these wobbling sweethearts. The region's a picturesque paradise.

SEE IT! *Car ferry MS Norröna (operated by Smyril Line) from Denmark sails right into Seyðisfjörður via the 17km-long (10-mile) fjord. Borgarfjörður Eystri is a 92km (57-mile) drive north from here.*

447

Explore wild beaches on horseback in Tayrona National Park

COLOMBIA // This could be the first time you've hired a horse to get to a campsite. Finding the ultimate beach – nestled in a curve of the Caribbean, between the foothills of the Sierra Nevada de Santa Marta and the coast – takes commitment. Once you've negotiated the jungle track, pitched your tent and strung up your hammock, it's time to chill with a book and a beer. Mind you, there's a sultry swathe of Caribbean coast to explore, across six sensational bays – Chengue, Gayraca, Cinto, Neguanje, Concha and Guachaquita. All have primo beachcombing and some are calm enough to snorkel. And there are rumours of archaeological remains in the jungle...maybe that book can wait.

SEE IT! *Take a bus from Santa Marta to the park entrance. Cañaveral is the closest and busiest beach. For solitude, saddle up and explore.*

448

Bask in the awesome presence of Cerro Aconcagua

ARGENTINA // Standing proud among the Andean Mountains that divide Argentina and Chile, awe-inspiring Aconcagua attracts climbers from all over the world. Those who manage to summit the tallest peak outside Asia can claim to have stood atop the 'roof of the Americas', but at nearly 7000m (22,965ft), the altitude and severe weather conditions make this a challenging climb.

It was in the shadow of Aconcagua that in 1817 General José de San Martín and his army crossed the Andes to help liberate Chile from the Spanish, and in 1985, climbers discovered an Incan mummy at 5300m on the mountain's southern face. Unless you're an experienced climber, the jagged, humpbacked peak of Aconcagua – particularly pretty in the pink glow of sunset – is best admired from a distance.

SEE IT! *Aconcagua is roughly 200km (125 miles) west of Mendoza. Reaching the summit requires a commitment of at least 15 days.*

© rocharibeiro / Shutterstock

449

Relish the visionary architecture and food stalls of Rotterdam's Markthal

THE NETHERLANDS // Roaming Rotterdam's streets feels like strolling through a gallery of modern architecture. Razed during WWII, it offered a blank canvas for a no-holds-barred building spree that's continued ever since. An eye-catching addition is the 2014-opened Markthal (market hall). Shaped like an enormous horseshoe, the 40m-high (130ft) building's apartments arc over a soaring interior space with a 11,000-sq-metre (118,400-sq-ft) mural (the world's largest artwork) featuring outsized fruit and vegetables, which is splayed across its glass walls and ceiling. Dozens of stalls sell cheeses, herring and waffles, plus international dishes for on-the-spot snacking. Behind-the-scenes tours can take you into the logistics centre, refrigeration rooms and one of the apartments, and introduce you to stallholders (with tastings too).

SEE IT! *Frequent trains run between Rotterdam and Amsterdam via Schiphol International Airport; there's a London Eurostar rail service.*

© Kiev.Victor / Shutterstock

450

Get mellow in the Blue Lagoon

MALTA // You don't need an Instagram filter to share pics of this sheltered cove – it couldn't possibly be bluer. On warm days, the cove's rocky ledges are draped with beachgoers, who tan languorously in the Mediterranean sun between splashing and snorkelling sessions. Strong swimmers can paddle across the lagoon's diamond-clear water to the uninhabited islet of Cominotto, with its sliver of sand as soft as almond flour.

SEE IT! *The lagoon is on the tiny island of Comino, accessible from Malta and Gozo by ferry or boat tour. Summer season is rammed.*

452

451

Toast the sunset from Belgrade's Kalemegdan Citadel

SERBIA // Looming over Belgrade, the 1500-year-old Kalemegdan is more than a scenic stronghold: it's a glorious time machine. Strategically positioned above the confluence of the Sava and Danube Rivers, it has been fought over by countless invaders over the centuries. The juxtaposition between its gory history – evident in still-standing prison gates, torture towers and a military museum – and today's boho cafes, tennis courts and chess players, makes Kalemegdan simply spellbinding.

SEE IT! *Use the audio guide and map to learn about the citadel's bloody history, or book a local tour to discover what lies beneath.*

452

View the big castle collection of little Luxembourg's prosperous capital

LUXEMBOURG // For such a tiny country, Luxembourg has a profusion of fortifications. In its capital, Luxembourg City, spectacularly set across plunging river gorges, the Bock Casemates fortress was raised by Count Siegfried in AD 983, with rock galleries and passages that were carved by the Spaniards in 1644. It was superseded over the centuries by a string of forts and castles tasked with protecting the Duchy's fabulous wealth. Today, wandering the castles is a tour through the ages.

SEE IT! *Take in the views from the Chemin de la Corniche, a 600m (1970ft) rampart promenade dubbed 'Europe's most beautiful balcony'.*

© Justin Foulkes / Lonely Planet

© aphotostory / Shutterstock

453

Find a head for heights at Jebel Shams' Wadi Ghul

OMAN // It's Oman's best view but you'll need a head for heights to appreciate it. At 3009m (9872ft), Jebel Shams is the country's highest peak. You're not here to bag the summit though – it's a military site so closed to the public. Instead, it's the plateau on the road to the top that's the star. This is the lip of the deep fissure of Wadi Ghul, where the canyon cliffs shoot down more than 1000m (3280ft). Unsurprisingly, it's known as Arabia's Grand Canyon. Various viewpoints lay along the rim offering vistas into and across the abyss. For extra action, and plenty of mountain goats for company, thrill-seekers hike the Balcony Trail along the cliffs, downwards to an abandoned village hewn into the rock below.

SEE IT! ***The Jebel Shams plateau is day-tripping distance from Nizwa but there are also hotels on the plateau itself.***

454

Step into a scroll painting atop Huangshan

CHINA // The ephemeral views of China's sacred Yellow Mountain have been captured by poets, painters and artists for thousands of years. These days at Huangshan you have to elbow in for views of cloud-dusted peaks and soaring canyons, and finding solitude for poetry writing is a tough ask. Legions of visitors come from across China and the world to see Huangshan's spectacular beauty, climbing its seemingly endless staircases for selfie-snapping opportunities. Not that we can blame them – Huangshan's mystique is undeniable. In order to capture a glimpse of the mountains the way the ancients experienced them, off-season is the best bet. Crowds thin in winter (apart from Chinese New Year) and you might just get a snow-sprinkled ledge to yourself.

SEE IT! ***Tunxi (Huangshan City) is an hour by bus from the peak and tourist village. Cable cars whisk non-hikers from hotels to peaks.***

455

Walk the laneways of an ancient Silk Road fortress in Ichon-Qala

UZBEKISTAN // The final resting stop for traders and caravans heading into the desert before they moved south to Persia, the city of Khiva was one of the Silk Road's most important oasis towns. Khiva's preserved and restored 'inner town', the Ichon-Qala is more than a fort – it's an entire city of mud-brick medressas, mosques and minarets situated along laneways and courtyards, much as it would have been 600 years ago. As with all of Uzbekistan's monuments, the Soviets went to great effort to restore and rebuild Khiva from disrepair, and today it is a drawcard Silk Road site.

SEE IT! *Khiva developed as it did for a reason – it is located on the edge of the vast Kyzylkum and Karakum deserts in far western Uzbekistan, and served as a key refuelling and resting point for Silk Road traders. Getting here requires a flight or an overnight train, although a high-speed rail line is under construction.*

© Sunrise Odyssey / 500px

© Liz Coughlan / Shutterstock

456

Follow humanity's story at the Louvre Abu Dhabi

UNITED ARAB EMIRATES // While modern-architecture fans may be content gawping at the Jean Nouvel–designed filigree dome that hovers over Abu Dhabi's new museum, it's what's inside that makes this a 21st-century cultural icon. The Louvre Abu Dhabi represents a revolution in museum curation – transcending borders, artefacts are clustered together by theme rather than specific civilisation or land. As you walk through the eras of human history from the Neolithic to today, equal billing is given to smaller, lesser-known cultures alongside pieces from well-known empires and countries. The result is a series of 12 galleries, with a collection that romps from a third millennium BC Bactrian princess statue to Ai Weiwei's *Fountain of Light*, which celebrate our interconnected, globalised world, highlighting the links and connections that have been forged between cultures for centuries.

SEE IT! *The museum is on Saadiyat Island in Abu Dhabi city. Evening visits avoid the tour buses from Dubai.*

457

Go with the flow in Si Phan Don

LAOS // Si Phan Don (Four Thousand Islands) is where Laos truly becomes the land of the lotus eaters: an archipelago of islands where the pendulum of time swings at half speed and life drifts by as lazily as the murky waters of the Mekong River. Some islands are backpacker playgrounds, but others are sleepy spits of sand dotted with tranquil fishing villages where you'll find inner and outer peace swinging in a hammock. Those who need more activity can pedal about on a bicycle, potter around local markets, take to the water in a kayak, spot dolphins and visit waterfalls (one of which you can zip line over). Don't miss a side trip to the Khon Phapheng Falls, a set of tumbling rapids that's the largest and most impressive of any on the Mekong.

SEE IT! ***Si Phan Don is in Champasak Province. Islands Don Det and Don Khon have the pick of the accommodation and activities.***

© S-F / Shutterstock

© ITAR-TASS News Agency / Alamy Stock Photo

458

Gaze on the marvels of Kelvingrove Art Gallery & Museum

SCOTLAND // Artistic, inquisitive, dressed in red sandstone and surrounded by a gorgeous park, Kelvingrove is Glasgow's best museum. And in a city where energy, culture and history bubble and fizz like freshly broached Irn-Bru, that's saying something. This grand Victorian cathedral of culture has an eccentric array of exhibits. You'll find fine art alongside stuffed animals, and Micronesian shark-tooth swords alongside a Spitfire plane – plus regular organ recitals. The mix can make your head spin, but rooms are thoughtfully themed, and if you want to focus on local highlights you can check out the rooms exploring the Charles Rennie Mackintosh and the Glasgow School. Other highlights include French impressionist and Renaissance paintings. Even Salvador Dalí has a spot here, with his superb *Christ of St John of the Cross*.

☛ SEE IT! ***Kelvingrove Art Gallery & Museum is in Glasgow's West End. Entry is free, and free hour-long guided tours run twice daily.***

459

Feel the chill of history at Perm-36

RUSSIA // Writers, artists and other freethinkers were imprisoned within the chilly walls of this Soviet labour camp between the 1940s and the 1980s. It's one of the only remaining Gulags in the former USSR, worth visiting despite recent government efforts to make it more 'patriotic'. Visitors in the past five years have seen exhibits revamped to remove descriptions of hardship and oppression, and to emphasise that the prisoners here were criminals (rather than political dissidents). Read some outside history before you arrive; it will make the bare wooden barracks, isolation cells and barbed wire fences more poignant. Even if you don't believe in ghosts, the entire place feels haunted, especially when the skies are the iron grey of winter and the Siberian winds howl.

☛ SEE IT! ***It's near Kuchino village, 25km (15 miles) from Chusovoy, which is 100km (62 miles) from Perm, in Russia's Ural Mountains.***

MINHO
PORTO
ENTRADA
Chegadas

© photoiasson / Getty Images

São Bento station's gorgeous azulejos (decorative tiles) feature battles, knights, weddings and scenes of rural life

460

Hit the tiles at Porto's beautiful São Bento station

PORTUGAL // Whisking you back to a more genteel age of rail travel, Porto's São Bento train station was designed by architect José Marques da Silva in the Beaux Arts style in 1903. The facade is pretty enough with its mansard roof, but step inside to understand what all the fuss is about: namely azulejos. Some 20,000 decorative tiles dance across every inch of wall space. These bear the imprint of master tile painter Jorge Colaço, who worked his magic on the station between 1905 and 1916.

Large-scale blue-and-white azulejo panels force you to gaze up to scenes depicting milestone events in Portuguese history: Prince Henry the Navigator's conquest of Ceuta in 1415, for instance, and the Battle of Valdevez between the Kingdom of León and Portugal in 1141.

SEE IT! *São Bento is on Praça Almeida Garrett in central Porto. To see the station at its peaceful best, overnight at the retro-cool Passenger Hostel.*

Step inside artworks at Paris' digital-art museum, L'Atelier des Lumières

FRANCE // Paris' creative nerve centre, the 11th arrondissement (city district), is the home of France's first museum for digital art, which opened in 2018. An early 19th-century foundry now forms the backdrop for dazzling multisensory exhibitions. Within this innovative space, you can take a virtual stroll through a changing array of world-renowned artworks as they're projected on the bare stone of its 1500-sq-metre (16,145-sq-ft) exhibition hall, accompanied by an atmospheric soundtrack, providing an immersive, almost meditative experience.

Programs have included the vivid creations of Gustav Klimt and the expressive brushstrokes and intense colours of Vincent Van Gogh's works swirling across the walls and creating a magic carpet effect on its floors, as well as the Zen-like 'Japan dreamed: images of the floating world'.

☛ SEE IT! ***It's not possible to buy tickets at the museum: pre-book them online.***

Inge Solheim's Top Five Places

Inge Solheim is one of the world's leading adventure guides, taking high-profile individuals into some of the most remote spots on the planet. He is best known for leading the Walking with the Wounded expeditions to the North and South Poles with Prince Harry. He is an ambassador for Breitling.

LOFOTEN, NORWAY – An amazing place where nature, culture and climate stimulate all your senses. The peaks and fjords contrast with beautiful beaches and fishing villages and there's hiking, surfing, skiing, fishing, birdwatching, kayaking and whale watching.

MOAB, UTAH – This is my favourite desert, with beautiful canyons, impressive rock formations, rivers and lakes. Head deep into the canyons and you can see old settlements, cave paintings and even dinosaur footprints.

ALMATY, KAZAKHSTAN – The first time I went, I fell in love with this city and the nature that surrounds it. In one day you can visit glaciers, high mountains, lakes, canyons, deserts with sand dunes and the never-ending steppes.

SVALBARD, NORWAY – This arctic archipelago is very special. Apart from a few settlements, most of Svalbard is wilderness. There are 2500 humans, 4000 polar bears, lots of reindeer, polar fox and millions of birds. The ocean is full of whales, seal, walrus and fish.

THE WORLD'S OCEANS – I've done everything from diving on Caribbean reefs to kayaking in Greenland. We are so dependent on healthy oceans and we have to stop polluting them with plastic, sewage, toxins and noise.

462

Feel the follies at gorgeous Castle Howard

ENGLAND // Castle Howard has 145 rooms, took a century to build and was big enough to have its own railway station until the 1950s. Thanks to its role in both the TV series and the film of Evelyn Waugh's *Brideshead Revisited*, for many it's the definitive aristocratic estate. To cap it all, it's still owned by the Howard family (Anne Boleyn and Catherine Howard are relations), who began building it in 1699 on the site of ruined Henderskelfe Castle.

The house took so long to build that it has two entirely different styles – exuberant baroque in the earlier East Wing, and simpler Palladian in the later West Wing. Inside are enormous riches: statues, an altar, the grand Great Hall, magnificent stained-glass windows and paintings by Van Dyck and Hans Holbein.

Yet the gardens are many visitors' favourites. It's a classic tamed landscape, with artificial lakes, rhododendrons, peacocks and rare trees adorning a great swathe of parkland. Classical statues and pyramids are joined by the Temple of the Four Winds, a folly that's half cutesy holiday lodge, half Roman palace. Wandering the grounds, you feel transported by an environment that mixes the honest pleasures of the outdoors with the extravagant fantasies of long-gone nobles.

SEE IT! ***Entry to the house is 10.30am to 4pm and the gardens 10am to 5pm. Direct buses run from York.***

463

Swashbuckle your way around colonial Old San Juan

PUERTO RICO // Imagine a Caribbean pirate's hideaway. The result will fall short of the reality of the splendid colonial heart of San Juan, Puerto Rico's capital. Shoehorned onto an islet guarding San Juan's harbour, and connected to the rest of Puerto Rico only by bridge, this history-steeped district feels surrounded by shimmering blue ocean. San Juan was among the first European settlements founded in the Americas (1521) and for a long time was one of the wealthiest too; the temptation was too much for ransacking pirates, including Sir Francis Drake. Old San Juan's key buildings are the bombastic forts of El Morro and San Cristóbal, the bulky bulwarks of which show the settlement's early need to repel invaders. Add a labyrinth of sinuous cobbled streets and the district's legendary drinking dens and you have one very comfortable enclave for a corsair. That said, it's not all swashbuckling swagger. Great galleries and museums, plus one of the Caribbean's best dining scenes, ensure Old San Juan caters to sophisticated sorts too.

SEE IT! ***Flights connect San Juan with several mainland US and European cities.***

© Ana Flašker / Alamy Stock Photo

© Tanatat Pongphibool, Thailand / Getty Images

464

Visit a modern masterpiece in Baku's Heydar Aliyev Center

AZERBAIJAN // The cornerstone of post-Soviet Baku, this Zaha Hadid-designed masterpiece is like a 21st-century Sydney Opera House that's been draped with a layer of fondant icing. Vast and impressively original, the building was designed to express the optimism of a nation. The voluminous interior is a celebration of light and space, with undulating shapes morphing into more familiar structures like stairways and balconies. The centre hosts temporary exhibitions, events and a museum laid out over three floors, recounting Azerbaijani history and the role of the building's namesake, Azerbaijan's president from 1993 to 2003. The greatest delight, though, comes from simply wandering around the extraordinary exterior and discovering endless perspectives to contemplate and photograph. You'll never get bored.

SEE IT! ***It's free to enter the building on the edge of central Baku, but the museum and exhibitions command a small fee.***

465

Touch the beating heart of Eurasia in Tbilisi Old Town

GEORGIA // A tangle of winding lanes, wooden houses and handsome churches overlooked by the 4th-century Narikala Fortress, Tbilisi Old Town is a charming corner of the Caucasus. The fortress – vying for domination of the skyline with the 20m-tall (65ft) Kartlis Deda (Mother Georgia) monument – owes its ruinous state to an explosion of Russian munitions in 1828. Once similarly dilapidated, a clutch of perilously tilting art nouveau houses in the streets below underwent a huge restoration in 2009 and the glow-up has drawn more visitors than the narrow lanes were ever designed for. Plenty more buildings remain in a state of picturesque but precarious disrepair, but they may not stay standing for long. Grab a steaming cheese bread from one of the busy cafes and soak up the scene while you can.

SEE IT! ***The Old Town's streets are choked with admirers in summer. Take the cable car up to Narikala for a bird's-eye view.***

466

Celebrate the rebirth of African wildlife in Akagera National Park

RWANDA // The Lazarus of African parks, Akagera National Park has seen a remarkable rebirth over the past decade and its success means Rwandan wildlife is no longer all about gorillas.

Head out onto its East African savannah at dawn in search of safari wildlife that once again includes the Big Five. Lions and rhinos were reintroduced in 2015 and 2017 respectively, and their populations, along with those of leopards, elephants and buffalo, are surging.

Solid sustainability management by the NGO African Parks has not only ensured healthy animal numbers and improved tourism facilities, but also increased community engagement. Some 1800 local students are invited into the park each year as part of an environment education programme, and solar-powered predator fences have reduced human-wildlife conflict. Impressively, now more than half of park visits are by Rwandan nationals.

☛ SEE IT! ***Safari companies in Kigali can arrange visits, but it's cheaper to simply rent your own 4WD there.***

467

Release your inner archaeology nerd at Pafos

CYPRUS // Sprawling Pafos Archaeological Site will fire the imagination of your inner Indiana Jones. It's a tantalising work in progress: what you see is only a modest part of the 4th-century-BC ancient city, with many treasures still to be unearthed here. Highlights include the mesmerising collection of intricate mosaics, whispering stories from ancient Greek myths, including the hero Theseus fighting the Minotaur and the tragic tale of Phaedra's love for her stepson.

SEE IT! *The site is located in Cyprus' southerly resort of Pafos; there's a free car park near the entrance west of Kato Pafos.*

468

Swim, dive or hike in the tropical Con Dao Islands

VIETNAM // With Thailand's beaches straining, shrewd travellers look elsewhere for adventurous getaways. The low-key Con Dao archipelago may just fit the bill. The main island, Con Son, has turquoise waters and coral reefs, a mountainous interior and forests for hikes and wildlife watching and a tiny capital for market breakfasts and strolls past French colonial-era villas. It's a pretty face with a complex history. Thousands of prisoners once languished – former prisons tell their story.

SEE IT! *Fly from Ho Chi Minh City, or take the (sometimes bumpy) 2½-hour boat ride from Tran De in the Mekong Delta.*

469

Tour the memorials to carnage in the Gallipoli cemeteries

TURKEY // Turkey's Gallipoli Peninsula is a place of remembrance for the WWI battles that took place on its rugged shores. The nine-month campaign (April 1915 to January 1916) resulted in half a million casualties, and the cemeteries and memorials scattered across the pine-clad countryside are a stark reminder of the brutality of war. Especially for visitors from Turkey, Australia and New Zealand, this is an important pilgrimage site. The thousands of graves, a testament to the sacrifice and heroism of the Allied and Ottoman soldiers, are given extra poignancy by the serene landscape.

SEE IT! *Most visitors base themselves in Eceabat or across the Dardanelles in Çanakkale.*

Remember the bitter past at the Museum of WWII

POLAND // The exhibits begin mildly enough, in the reconstructed 1930s apartment of a family of Polish Jews. But the sense of horror quickly grows. Nazi propaganda posters. Sherman tanks. A hall lined with images of bombed-out cities. Then, most affectingly, personal artefacts – farewell notes, house keys belonging to Jews who never came home. Ironically, this masterwork of remembrance has been under attack since it opened in 2017, by a right-wing government that says it's not sufficiently patriotic.

☛ SEE IT! *The stark, modernist museum is in Gdańsk. Buy tickets ahead online as there's an hourly visitor limit.*

Find your perfect tropical retreat in the Mamanucas

FIJI // The tourist board have it easy on Fiji's Mamanucas, an island chain where beautiful palm-fringed beaches overlook brilliant blue seas. These islands are deservedly Fiji's most popular visitor destination, and they're spread out enough to offer something to everyone without ever feeling too crowded, whether it's a family-friendly resort or stripped-back beach huts. Water sports abound, whether you're a surfer looking for a break, a chilled-out paddleboarder, or a diver itching to explore the kaleidoscopic coral reefs below the waves.

☛ SEE IT! *Catamarans, speedboats and a seaplane service link the Mamanuca Islands to nearby Nadi, home to Fiji's main airport.*

Marvel at a prototype of modernism at Villa Tugendhat

CZECH REPUBLIC // This pioneering Czech edifice is the work of German architects Ludwig Mies van der Rohe and Lilly Reich and is considered one of the first truly modernist buildings anywhere in the world. Sequestered away in the wealthy Brno suburb of Černá Pole, the residence was commissioned by Jewish Germans Fritz and Greta Tugendhat, who lived here until being forced to flee Czechoslovakia in 1938. The groundbreaking result was one of the first open-plan houses and is a shrine for devotees of modern architecture to this day.

☛ SEE IT! *From Brno, the biggest city in the eastern Czech Republic, take tram 5 to Černopolní, from where the villa is a short walk.*

Feel dwarfed by an earlobe at the Le Shan Grand Buddha

CHINA // For such a humble man, the Buddha has been reproduced on a grand scale over the centuries. The seated Buddha statue at Le Shan in Sichuan province has the ancient teacher blown up to 40 times normal size, with 7m-long (23ft) earlobes and a 6m-long (20ft) nose. In fact, this vast representation, created in the 8th century by monks, is less a statue than a carved mountain. To really grasp the scale of the Grand Buddha, climb the stairway to his right shoulder for a close-up view of one enormous earhole.

☛ SEE IT! *Ferries run from the main dock in Le Shan; it's more atmospheric than coming by road. Afternoons are less busy.*

© K3S / Shutterstock

A water-spouting giant guards the entrance to Swarovski Kristallwelten – a sparkling fantasy world in the Tyrol

474

Be spellbound by water giants and crystal trees at Swarovski Kristallwelten

AUSTRIA // At first glance Wattens seems like any other little Tyrolean village. But brace yourself for bling: here is Swarovski Kristallwelten, a sparkly-spangly world in which the famous crystal creators have unleashed the full force of their fantasy. A water-spouting giant welcomes you to the park, where the *Crystal Cloud*, bejewelled with 800,000 hand-mounted crystals, hovers above the glittering *Mirror Pool*.

The Chambers of Wonder dazzle with such showstoppers as *Silent Light* by Alexander McQueen and Tord Boontje, a Narnia-like winterscape of snowflakes and crystal trees, and South Korean artist Lee Bul's *Into Lattice Sun*, a mind-bending crystal and mirror installation. Be sure to check out Brian Eno's geodesic *Crystal Dome*, where music and mirrors combine to extraordinary effect.

SEE IT! ***Wattens is in the Tyrolean Alps, 17km (11 miles) east of Innsbruck. Regular buses run from Innsbruck's train station to Kristallwelten (a 30-minute ride).***

Image © Elena Pominova / Shutterstock. Building © Cité du Vin / XTU architects

Bordeaux's Cité du Vin explores wine regions around the world and has a rooftop bar with views of the local vineyards

175

Explore the world of wine at Bordeaux's Cité du Vin

FRANCE // Bordeaux has long been famed for the vineyards ribboning the surrounding countryside, but it wasn't until 2016 that the city unveiled a major museum dedicated to its signature drop.

Rising above the banks of the River Garonne near Bordeaux's industrial port, the Cité du Vin (City of Wine) is housed inside a glinting gold building that resembles a gigantic wine decanter. Its permanent exhibition has six themed areas with interactive stations that include a virtual flyover of wine regions in 17 countries, an explanation of winemaking and its history and sniffing out varietals. Wine-related temporary exhibitions are also staged at the museum. More interactive still, you can sip wine at one of the museum's many wine-tasting ateliers (workshops).

After your visit, you'll receive a free glass of wine in the panoramic all-glass rooftop bar Le Belvédère. A highlight in every sense, it has a 30m-long (100ft) bar, a ceiling made from thousands of wine bottles, and views of the river, city and vineyards beyond.

SEE IT! *From central Bordeaux, it's a scenic 2.5km (1.5-mile) stroll north along the river; alternatively, take Tram B from Esplanade des Quinconnes to the museum's own tram stop.*

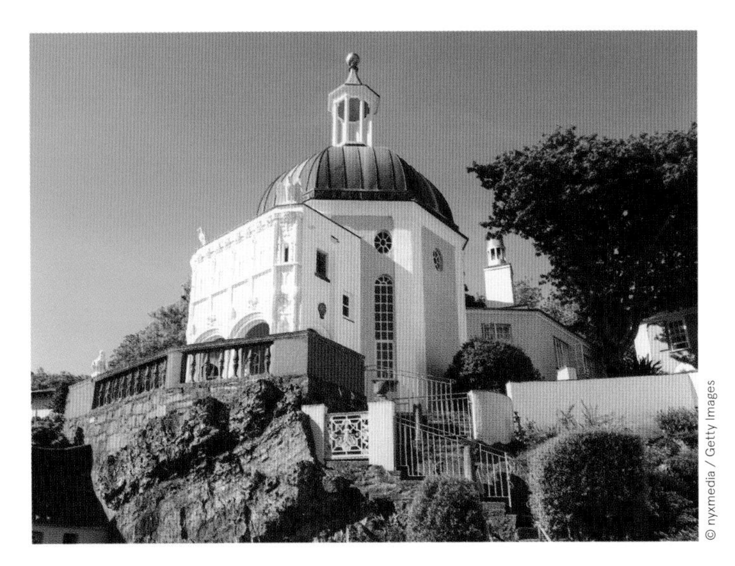

© nyxmedia / Getty Images

© Rob van Esch / Shutterstock

476

Enter the potty world of *The Prisoner* at Portmeirion

WALES // If you've ever watched the kooky 1960s TV series *The Prisoner*, you might find yourself looking over your shoulder for giant bouncing balls as you explore curious Portmeirion in Wales, where the cult show was filmed. This collection of colourful Italianate buildings overlooking a vast sandy beach was created by Welsh architect Sir Clough Williams-Ellis, who lived by the maxim that beauty in life is a necessity. It was also the former home of the Portmeirion pottery factory, whose florid tableware graces untold British dining tables. It's all rather captivating and slightly unnerving, given this is a village with no permanent residents. Something you would find nowhere else but in Britain, for sure.

SEE IT! ***Portmeirion is 5km (3 miles) southeast of Porthmadog, on the Shrewsbury-Pwllheli Cambrian Coast Railway; public transport isn't great, so if you don't fancy the walk, catch a taxi.***

477

Find fine wines and Gothic churches in the medieval maze of Porto's Ribeira

PORTUGAL // Prepare to be smitten by this World Heritage maze of alleyways, rising up from the Douro River in a helter-skelter of pastel-coloured houses, bell towers and exuberantly gilded baroque and Gothic churches. Cobbled lanes twist past pavement cafes and craft shops, delis and tempting little wine bars, while a flight of seemingly never-ending steps leads to the city's crowning glory, the fortress-like Sé cathedral. Down by the river, you can sense the Atlantic as gulls squawk overhead, sardines sizzle and traditional wooden *rabelo* boats putter by. Here the double-decker arched Ponte de Dom Luís I makes a spectacular leap across the Douro to Vila Nova de Gaia, where historic port-wine lodges offer tastings of prized rubies and tawnies.

SEE IT! ***The teleférico from Vila Nova de Gaia offers an excellent aerial view. On 23 June, there are festivities and fireworks for Festa de São João.***

© Juan Barreto / Getty Images

© Hiromi Kano / 500px

478

Soar above Bogotá on the TransMiCable

COLOMBIA // High in the Andean Altiplano, Bogotá's sprawl climbs the mountains as far as the eye can see. The hilliness makes commuting tricky, especially for residents of the city's disadvantaged south side. But a brand-new cable car aims to connect locals – and vista-seeking visitors – from the southern Ciudad Bolívar district to the city's bus system. Dubbed the TransMiCable, the apple-red cable cars float 7200 passengers an hour above a patchwork of parks, plazas and pastel bungalows. You'll be treated to views across the urban plateau from mountaintop to mountaintop. At night, the city is as spangled with lights as the Milky Way. The 15-minute ride can shave hours off daily commutes for southern residents. Best of all, the solar-powered gondolas have almost no carbon footprint.

☛ SEE IT! ***The TransMiCable connects Ciudad Bolívar with four TransMilenio bus stations: Tunal, Juan Pablo II, Manitas and Mirador del Paraíso.***

479

Hike a new Pacific path on the Michinoku Trail

JAPAN // An impressive newcomer on the long-distance hiking scene, the Michinoku Coastal Trail was born out of disaster – an initiative to promote recovery in the communities of Tōhoku (northeast Japan) that were devastated by the 2011 earthquake and tsunami. The roughly 700km (435-mile) trail snakes from Hachinohe in Aomori Prefecture to Sōma in Fukushima Prefecture, through some of the country's most scenic regions. Hikers traverse pine forests and fields of wildflowers, cliff-backed coves and gritty fishing villages, with mountain climbs and boat trips along the way. Don't have a couple of months to spare? Take your pick of 28 varied sections, which include day and half-day walks. The MCT is also still relatively unknown. While queues gather to summit Mt Fuji in peak season, visitors here can enjoy crowd-free rambles.

☛ SEE IT! ***Hachinohe is three hours from Tokyo on the bullet train, which also stops at other trail access cities, such as Sendai and Ichinoseki.***

480

Discover a design for life at Dessau's Bauhaus Museum

GERMANY // 'Form follows function' was the mantra of such early Bauhaus pioneers as Walter Gropius, who spearheaded the radical movement in 1919 in an effort to unite art, architecture and industry. With strong, clean lines and a 'less is more' ethos, the globally influential school flourished in Dessau. Just over a century later, you can feel its reverberations at the Bauhaus Museum, a slick glass-and-steel rectangle that shelters an impressive 1000-strong collection.

SEE IT! *Dessau is in Germany's Saxony-Anhalt state. Tickets are timed (admission is valid for an hour). Guided tours run at 1pm on Wednesdays and Saturdays..*

© Matthias Wehnert / Alamy Stock Photo

481

Wade into the watery wilderness of Everglades National Park

USA // Called the 'River of Grass' by its initial Native American inhabitants, Florida's Everglades is not just a wetland, or a swamp, or a lake, river, prairie or grassland – it is all of the above, twisted together into a series of soft horizons and enormous sunsets. The park's 64,238 hectares (1.5 million acres) provide ample occasions to spy alligators basking in the noonday sun, stroll boardwalks amid the birdsong and kayak in mangrove canals where a manatee might pop up.

SEE IT! *Everglades National Park has three entrances – Ernest Coe, Shark Valley and Gulf Coast – all accessible from Miami and best reached by car. Prime wildlife viewing time is December to March.*

© Justin Foulkes / Lonely Planet

Opposite, from top: Blooming bougainvillea in the pretty Tunisia Sidi Bou Saïd. Bottom: fresco-plastered Rila Monastery dates from the 10th century

482

Gaze on Tunisia's true colours in Sidi Bou Saïd

TUNISIA // With its stunning whitewashed buildings punctuated with bright blue doors and windows, the clifftop village of Sidi Bou Saïd looks every bit the perfect Mediterranean painting. No surprise, then, that painter Paul Klee, author André Gide, philosopher Michel Foucault, and arty British eccentrics Osbert and Edith Sitwell once wandered these cobbled streets. Lush palm trees stand to attention, bougainvillea flowers cascade over gates and through arches, and striped cafe cushions soak up the sun, the hazy purple hills of Cap Bon just visible in the distance. Sidi Bou Saïd's winding, hilly streets invite further exploration: galleries, *dars* (traditional guest houses), rooftop cafes and gourmet restaurants hide behind demure signs. Below the village, the shimmering waters of the Med beckon. Many cafes charge more for the view than the quality of their food and drink, but when golden hour arrives, you'll be happy to have shelled out for a front-row seat.

SEE IT! ***Sidi Bou Saïd is northeast of central Tunis and is easily reached on the suburban TGM train from Tunis Marine station.***

483

Peek up at the divine delights of Rila Monastery

BULGARIA // Shrouded in both forest and history, the remote Unesco-listed Rila Monastery is a sacred, sylvan symbol of Bulgarian identity. Emerging from the mysterious Rila Mountains like a divine vision, the comely cloisters (established AD 927) were a stronghold of Bulgarian culture and language during Ottoman rule. If the mountain views, riotous fresco-plastered interiors and the 220-year-old Rila Cross – a wooden masterpiece of painstakingly carved biblical scenes so tiny they cost the artist his sight – don't leave you breathless, a hike up to the cave-tomb of St Ivan, founder of the monastery, and Bulgaria's very first hermit, will certainly do the trick. It is truly spectacular terrain up here, with the Rila Monastery Nature Park that surrounds the monastery soaring up to 2700m (8860ft) in a mythical mix of peaks, alpine meadows and forests: you may just find that, like St Ivan, you end up spending much more time in this beguiling area.

SEE IT! ***Rila is an easy day trip from Sofia (2½ hours), or stay overnight in a spartan monk's cell.***

482

483

484

Call on polar bears and beluga whales in Churchill

CANADA // No road goes to this wee town on the tundra beyond the tree zone, at the very edge of Hudson Bay. Only a lengthy train or plane ride will get you to Churchill. It's worth the effort, though, for the extraordinary experiences that await. First up: polar bear spotting. Churchill, as it happens, is right in the bears' migration path. About 900 of the snowy beasts hang out here, waiting for the bay to freeze and hunting season to begin. Tundra buggies head out in search of the animals, which sometimes prowl close enough for you to lock eyes. Then there are beluga whales. Some 3000 of the glossy white creatures summer in Churchill River, and you can kayak among the chirpy pods. Not wowed enough yet? The Northern Lights let loose with shimmering green-yellow abandon from October to March. Add in the town's Inuit art and culture, its lonely forts and Cold War missile sites and its backdrop of endless subarctic majesty, and you've got a destination like no other.

SEE IT! *Most visitors to Churchill depart via plane or train from Winnipeg, Manitoba. Peak polar bear viewing is mid-October to mid-November; peak whale watching comes in July and August.*

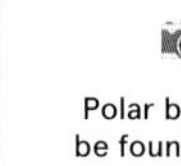

Polar bears can be found around Churchill in large numbers in October and November, before they head out onto the frozen bay in winter

© Rui Vale Sousa / Shutterstock

© gbarm / Getty Images

485

Ponder the beauty of Roman engineering at Segovia's aqueduct

SPAIN // It's a testament to Roman design aesthetics that their greatest engineering projects double up as handsome pieces of architecture. Contemporary Segovia, like any modern city, gets its water from a web of underground pipes, but in the 1st century AD it was dispensed via an elegant aqueduct of single and double arches that rose 28m (92ft) above the surrounding townscape. Built to last, using over 20,000 granite blocks but not a drop of mortar, the Acueducto still dominates Segovia's city centre, its formidable arches standing as a lasting reminder of Roman ingenuity. So enduring was the stone watercourse that it continued to supply the city with water until well into the 19th century, 1800 years after it was built.

SEE IT! ***Fast Madrid trains take less than 30 minutes. Climb the stairs from behind the tourist office for a different perspective.***

486

See the pyramids without the crowds at peaceful Dahshur

EGYPT // Home of Egypt's first true pyramid, Dahshur is just as impressive as Giza but far more peaceful. The two main pyramids on the desert plains here were built by Pharaoh Sneferu, father of Khufu, who went on to raise Giza's Great Pyramid. Sneferu's Red Pyramid (named for the hue of its weathered limestone) has long been the highlight, with a steep claustrophobic shaft leading down into the burial chamber. Until recently, you could only admire Sneferu's nearby Bent Pyramid (with its wonky sides that incline first at 54 and then 43 degrees) from outside, but in mid-2019 Egypt opened up the interior to visitors for the first time. Now you get to clamber into two pyramids without the Giza crowds.

SEE IT! ***Dahshur is best visited as part of a day trip from Cairo that includes the sites of Saqqara, 11km (7 miles) to the north.***

© saiko3p / Getty Images

© canadastock / Shutterstock

187

Ride Mi Teleférico, La Paz's 'subway in the sky'

BOLIVIA // The most futuristic way to get around the world's highest-altitude capital is via this series of cable cars, which bobble far above the terracotta roofs of La Paz. Use them to check out the psychedelic 'New Andean' mansions of the adjacent city of El Alto, which incorporate indigenous designs and patterns in their exteriors. Then sail on down to the Zona Sur for some bohemian boutique shopping and a trendy dinner.

SEE IT! ***There are 10 lines in service, and one pending. They run from dawn until 11pm (9pm on Sunday).***

188

Trace the Danube through Austria's romantic Wachau

AUSTRIA // Where the Danube gracefully curves through wooded hills, past apricot orchards and terraced vineyards, the World Heritage-listed Wachau presents an irresistible vision of romantic Austria. Come to hike, bike or winery-hop your way through the valley that inspired Strauss' 'Blue Danube' waltz. Don't miss the Benedictine abbey-fortress of Melk, with its baroque-gone-mad church, library and Marble Hall, vine-swathed Spitz, and Dürnstein with its eyrie-like castle ruins.

SEE IT! ***The Wachau is the most scenic stretch of the Danube Valley, an hour west of Vienna. Trains run to Melk and Krems an der Donau.***

© Nataliya Hora / Shutterstock

© Russ Heinl / Shutterstock

489

Step into Africa's ancient Punic capital of Carthage

TUNISIA // The remnants of grand Carthage lie scattered across a quiet seaside suburb northeast of Tunis. The Romans defeated the Phoenicians and took Carthage in the Third Punic War, using these same stones to reconstruct their own outpost. A two-storey museum and the excavated residential quarter atop Byrsa Hill offer a glimpse at its life. A series of Roman villas hint at the opulence of the new neighbourhood, with columned porticos, mosaic floors and terraces with views of the Med. The evocative seaside Antonine Baths are perfectly positioned a short walk down the hill – they were destroyed by the Vandals in AD 439. The scant remains of two superpowers that grew fat from trade are an enigmatic peek into how empires can crumble as quickly as they rise.

SEE IT! ***Carthage can be easily reached using Tunis' TGM suburban trains, which run from the central Tunis Marine station.***

490

Appreciate First Nations ingenuity at Head-Smashed-In Buffalo Jump

CANADA // The name paints the picture for this southern Alberta site. Buffalo Jump? Yep. For thousands of years, the Blackfoot people used the cliffs near Fort Macleod to hunt bison. Braves from the tribe would herd the animals towards the precipice and make them leap over it to their doom, thus providing meat for the people. Look around today and you'll see a lonely sweep of prairie under big skies, with a trail that leads to the spot where the buffalo plummeted. And Head-Smashed-In? That comes from a legend. A young brave wanted to watch the buffalo tumbling past from below, but the animals piled up so fast they trapped him. The weight of the carcasses crushed his skull. Blackfoot guides tell the story and explain the site's sacred nature.

SEE IT! ***Fort Macleod is the nearest town. It's also easy to reach from big cities Calgary and Lethbridge. The site is open year round.***

491

Hear the trumpets blow at Rynek Główny

POLAND // Nearly 800 years old, Rynek Główny – Kraków's bustling main square – is paved with cobblestones and surrounded by original buildings, with the old marketplace at its centre. A window opens in the tower of the Mariacka Basilica, and a trumpeter begins to play the mournful melody of the hourly hejnał. Far below, the crowds pause to listen. Suddenly, the warning song is cut short, mid-note – as if the trumpeter were shot in the throat by a Mongol invader – in a re-enactment of a legend that's also nearly 800 years old.

SEE IT! ***Kraków is stunning any time of year, but it can be bitterly cold from December to March.***

Click like crazy on picture-perfect Sveti Stefan

MONTENEGRO // The impossibly picturesque walled island village of Sveti Stefan guarantees the greatest 'wow' moment along Montenegro's Adriatic coast. Terracotta-roofed graceful stone villas and clusters of pine and oleander overlook a gorgeous pink-sand beach lapped by turquoise waters. Now converted into a luxury resort and off limits to non-guests, the island is linked to the mainland by a narrow causeway and can be visited on exclusive guided tours.

SEE IT! ***Sveti Stefan is also the name of the small town directly opposite the island. You can enjoy the views from its slopes.***

Play man on the moon at Lac Abbé

DJIBOUTI // This 'slice of the moon on the crust of the Earth' (as it's often called), actually has a lot more going on than Tranquillity Base. Walk across its parched, barren surface in the shadows of towering, steam-belching limestone chimneys, visit its mineral-rich hot springs, which sustain enigmatic local nomads, and admire pink flamingos posing against the rich blue sky. Stay for the unbelievable sunset, then spend the night to witness the sci-fi scene all over again at sunrise.

SEE IT! ***You'll need to rent a 4WD with driver in Djibouti City or arrange a trip with a tour operator.***

494

Get your desert thrills in Arikok National Park

ARUBA // An undulating desert-scape brims with electric-blue lizards and akimbo-armed cactuses, set against dramatic cliffs and a churning sea. Almost a fifth of the Dutch Caribbean island of Aruba is reserved for this startling beauty of a national park, and its lurid, dusty stretches are the highlight of any Aruban getaway. True adventurers should dunk themselves in the Natural Pool, a limestone, fish-frequented swimming hole over which massive waves crash. And the panoramic view from Mt Jamanota is a stunner.

SEE IT! ***Island tour operators offer four-wheeler tours galore, but you'll tread more gently on this ecosystem via horseback or on foot.***

Top: built in 1908, Shackleton's Hut has barely been touched since. Bottom: hundred-year-old socks still hang on its washing lines

495

Slip into a polar explorer's shoes at Shackleton's Hut

ANTARCTICA // Out on its lonesome at rocky Cape Royds in Antarctica, this wooden hut erected by explorer Ernest Shackleton in February 1908 has been preserved in ice – literally. Hardly a stick of firewood or can of rations has been moved in the intervening century; those socks dangling from the line were left there by members of the Nimrod expedition who must have departed in a hurry. Shackleton didn't make it to the pole, but he did leave behind this astonishing time capsule. Only in the deep-freeze Antarctic could such a relic be so perfectly preserved, from the freeze-dried buckwheat pancake that still lies in a cast-iron skillet to the bench piled with mitts and shoes.

SEE IT! ***Cape Royds is protected, and you'll need to join a tour; only eight people are permitted in the hut at one time.***

© Nicram Sabod / Shutterstock

Top: Kawarau Gorge Suspension Bridge was the world's first commercial bungee jumping site. Bottom: one, two, three – go!

496

Bungee jump in the sport's centre of Queenstown

NEW ZEALAND // At some point in history, someone decided jumping off a very high point with an elastic band tied to their feet was a good idea. Thanks to that person (reportedly a British adrenaline junkie in the late 1970s, though people from Vanuatu's Pentecost Island have been doing something similar, with vines, for centuries), bungee jumping now seems like a semi-reasonable thing to do. If you're going to take the plunge, do it in Queenstown, where the sport was first commercialised and where the pre-bungee views are some of the world's most striking. Leap from the adventure hub's most famous spot, the 1880s Kawarau Gorge Suspension Bridge, where your hair will practically touch the jade waters of the Kawarau River.

© Scott Fairchild / 500px

SEE IT! *Fly into Queenstown from across New Zealand and Australia. AJ Hackett Bungy (as it's spelled in NZ) is the world's oldest outfitter.*

497

Feast your senses at Lyon's Cité de la Gastronomie

FRANCE // France's third-largest city, Lyon, is its culinary heart, thanks to its bountiful produce, chequered-tableclothed *bouchons* (small bistros) and galaxy of Michelin-starred restaurants. Now, the Cité de la Gastronomie (City of Gastronomy) is its showcase. Opened in 2019 in the magnificent Hôtel-Dieu de Lyon, a 16th-century landmark on the Rhône river, its four floors stage exhibitions on the past, present and future of French cuisine. You'll also find food-and-wine workshops, tastings and demonstrations, as well as a restaurant, cafe and shop.

SEE IT! ***It's open daily, and is located on Presqu'île peninsula between the Rhône and Saône rivers, near Bellecour metro station.***

498

Make a pilgrimage to the Golden Circle's original geyser

ICELAND // Geysir is part of Iceland's 'Golden Circle' of natural wonders. Gullfoss, another point on this popular route, gets all of the waterfall glory, while Þingvellir, the third stop, is a marvelous rift valley. In this case, Geysir is the highlight for the explosion-lovers. That's right. If you want to see boiling hot water erupt dramatically from the ground, spurting as high as 70m into the air, this is the place for you. There's no danger, just the chance of getting misted, it's been happening for quite some time now, around 10,000 years. Always impressive, though.

SEE IT! ***In summer, regular tour buses from Reykjavik cater for day-trippers. Or hire your own car and make it part of a wider adventure.***

© David Lazar / Getty Images

Sail past floating villages and gardens on Inle Lake

MYANMAR (BURMA) // The surface of serene Inle Lake resembles a shimmering silver sheet, dotted with stilt-house villages, island Buddhist temples and floating gardens of fruit and vegetables. Exploring this aquatic wonderland by motorboat or traditional skiff is one of Southeast Asia's great pleasures. When eventually you do hit land, you'll encounter whitewashed stupas and Shan, Pa-O, Taung Yo, Danu, Kayah and Danaw tribal people at the markets that hopscotch around the lake on a five-day cycle.

SEE IT! ***The nearest airport to Inle is Heho, 40km (25 miles) northwest of Nyaungshwe, located at the north end of the lake.***

Tajik National Park is a peaceful place, but its landscape is packed with the drama of jagged peaks, deep lakes and high-altitude deserts

© Iryna Hromotska / Shutterstock

500

Stand on the roof of the world in Tajik National Park

TAJIKISTAN // This sparsely populated wilderness of 2.5 million hectares encompasses the Pamirs, a jagged Central Asian landscape of alpine deserts, deep lakes and glacial valleys that make up the world's third-highest mountain ecosystem. The epic tectonic forces that pushed up these hulking massifs, along with the world's other highest ranges – the Himalaya and Karakoram – radiate from right here, at a point called the 'Pamir Knot'.

Awe-inspiring geography aside, Tajik National Park is a tranquil land, where Siberian ibex and Marco Polo sheep graze, snow leopards roam and Pamiri villages lie deep within dramatic rocky valleys. Those who venture here bed down in homestays or yurts (most of the tourist accommodation is located in the towns of Badakhshan, Khorog and Murgab). The area's local nickname is Bam-i-Dunya (Persian for 'roof of the world') and, with a plateau above 3000m (9840ft), Tajik National Park will quite literally take your breath away.

SEE IT! ***To visit the Pamirs in eastern Tajikistan you need a GBAO travel permit, which you can apply for along with your visa. Public transport is scarce; a trip here requires planning and a vehicle.***

© Christian Kober / robertharding / Alamy Stock Photo

Second Edition
Published in August 2020
Lonely Planet Global Limited
CRN 554153
www.lonelyplanet.com
ISBN 978 1 78868 913 7

Printed in China
10 9 8 7 6 5 4 3 2 1

Publishing Director Piers Pickard
Associate Publisher Robin Barton
Commissioning Editor Dora Ball
Art Director Daniel Di Paolo
Layout Jo Dovey
Picture Research Katy Murenu
Editors Bridget Blair, Monica Woods
Print Production Nigel Longuet
Proofreading Karyn Noble
Cartography David Connolly, Wayne Murphy
Thanks to Sophie Dening, Flora MacQueen, James Smart, Polly Thomas, Anna Tyler, Regina Wolek, Yolanda Zappaterra
Front cover depicts the Santuário de Cristo Rei by Francisco Franco de Sousa

Written by: Isabel Albiston, Alexis Averbuck, James Bainbridge, Joe Bindloss, Paul Clammer, Laura Crawford, Megan Eaves, Mary Fitzpatrick, Gemma Graham, Paula Hardy, Ashley Harrell, Anita Isalska, Lauren Keith, Jessica Lee, Catherine Le Nevez, Emily Matchar, Bradley Mayhew, Carolyn McCarthy, Lorna Parkes, Matt Phillips, James Smart, Simon Richmond, Brendan Sainsbury, Brana Vladisavljevic, Kerry Walker, Luke Waterson, Nicola Williams, Karla Zimmerman.

Top Five interviews by Lizzie Pook

Lonely Planet offices

AUSTRALIA
The Malt Store, Level 3, 551 Swanston Street, Carlton Victoria 3053 Phone 03 8379 8000

IRELAND
Digital Depot, Roe Lane (off Thomas St), Digital Hub, Dublin 8, D08 TCV4

USA
Suite 208, 155 Filbert St, Oakland, CA 94607 Phone 510 250 6400

UNITED KINGDOM
240 Blackfriars Road, London SE1 8NW Phone 020 3771 5100

STAY IN TOUCH
lonelyplanet.com/contact

Paper in this book is certified against the Forest Stewardship Council™ standards. FSC™ promotes environmentally responsible, socially beneficial and economically viable management of the world's forests.